Sorption - New Perspectives and Applications

Edited by Karmen Margeta and Anamarija Farkaš

Published in London, United Kingdom

Sorption - New Perspectives and Applications
http://dx.doi.org/10.5772/intechopen.1000381
Edited by Karmen Margeta and Anamarija Farkaš

Contributors
Abdelaziz Bacaoui, Aleksandar Bojić, Alfredo Campos-Trujillo, Amri Afaf, Andrés A. Abin-Bazaine, Chinedu T. Egbosiuba, Chinmay Sadadekar, Danijela Bojić, Deven Kadam, Dragana Marković Nikolić, Ezeh E. Mbamalu, Garmes Hocine, Gerardo Buelna, German Antonio Giraldo-Giraldo, Goran Nikolić, Himangi Neve, Ijeoma J. Ani, Juan Carlos Lucas-Aguirre, Ljiljana Stanojević, Ljubiša Nikolić, Mabrouki Jamal, Mario A. Olmos-Marquez, Miloš Durmišević, Miloš Kostić, Misael Cortés-Rodríguez, Moufti Ahmed, Mountadar Mohammed, Najat Kamal, Nataša Simonović, Prasad Wadekar, Rosa Galvez, Siraj Bhatkar, Taoufik Mohamed, Uduak G. Akpan, Vinayak Wadgaonkar

First published in London, United Kingdom, 2025 by IntechOpen
IntechOpen is the global imprint of INTECHOPEN LIMITED, registered in England and Wales, registration number: 11086078, 167-169 Great Portland Street, London, W1W 5PF, United Kingdom

For EU product safety concerns: IN TECH d.o.o., Prolaz Marije Krucifikse Kozulić 3, 51000 Rijeka, Croatia, info@intechopen.com or visit our website at intechopen.com.

British Library Cataloguing-in-Publication Data
A catalogue record for this book is available from the British Library

Sorption - New Perspectives and Applications
Edited by Karmen Margeta and Anamarija Farkaš
p. cm.
Print ISBN 978-0-85466-644-7
Online ISBN 978-0-85466-643-0
eBook (PDF) ISBN 978-0-85466-645-4

If disposing of this product, please recycle the paper responsibly.

Meet the editors

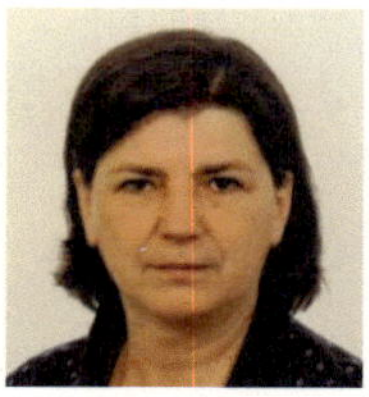

Dr. Karmen Margeta is a scientific advisor in the field of basic technical sciences. Her research focuses on the characterization of natural sorption materials using advanced analytical techniques and applying these materials in water treatment. She has been engaged in applied research related to material science for energy technologies. She has published more than eighty scientific papers. She is the co-author of publications that develop new paradigms and propose a new policy for the fight against climate change and sustainability for cities worldwide. Dr. Margeta invented the "Seawater Steam Engine" technology, which is presented as a potential solution to combat climate change by generating energy from seawater using renewable energy sources.

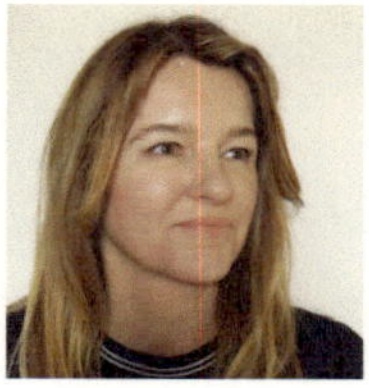

Dr. Anamarija Farkaš is a senior researcher at the Institute for Development and International Relations (IRMO) in the Department for Resource Economics, Environmental Protection and Regional Development in Zagreb, Croatia. She obtained a master's degree and a Ph.D. in Environmental Chemistry from the Faculty of Chemical Engineering and Technology at the University of Zagreb in Croatia. Her research interests include bioeconomy (blue economy), environmental protection, ecological agriculture, wastewater treatment, and forestry protection. She has participated in various multidisciplinary international and domestic projects. Dr. Farkaš is the author of numerous articles and studies published at scientific conferences, symposiums and workshops.

Contents

Preface **IX**

Section 1
Modeling of Sorption Processes **1**

Chapter 1 **3**
Defluoridation of Contaminated Mine Water by Adsorption on the Fly Ash in the Static and Dynamic Reactor: A Case Study in Morocco
by Moufti Ahmed, Taoufik Mohamed, Amri Afaf, Mabrouki Jamal, Mountadar Mohammed and Garmes Hocine

Chapter 2 **23**
Sorption of Phenolic Compounds from Woodwaste Leachate by Peat Media
by Najat Kamal, Rosa Galvez, Gerardo Buelna and Abdelaziz Bacaoui

Chapter 3 **53**
A Fixed-Bed Column Sorption: Breakthrough Curves Modeling
by Andrés A. Abin-Bazaine, Mario A. Olmos-Marquez and Alfredo Campos-Trujillo

Chapter 4 **73**
Thermodynamic Properties of Moisture Sorption and Glass Transition of Coconut (*Cocos Nucifera* L.) Powder Fortified with Physiologically Active Components
by Juan Carlos Lucas-Aguirre, German Antonio Giraldo-Giraldo and Misael Cortés-Rodríguez

Section 2
Novel High-Efficiency Sorbents **99**

Chapter 5 **101**
Bottle Gourd (*Lagenaria vulgaris*) Shell as a Natural, Biodegradable, Highly Available, Cheap, Agricultural by-Product, Miscellaneous Biomass, Ion Exchanger, Biosorbent and Fertilizer
by Goran Nikolić, Dragana Marković Nikolić, Aleksandar Bojić, Danijela Bojić, Ljubiša Nikolić, Ljiljana Stanojević, Miloš Durmišević, Nataša Simonović and Miloš Kostić

Section 3
Sorption Processes for Environmental Protection and Pollution Control **133**

Chapter 6 **135**
The Impact of Green Technology on Sorption Processes
by Ijeoma J. Ani, Uduak G. Akpan, Ezeh E. Mbamalu and Chinedu T. Egbosiuba

Chapter 7 **151**
Carbon Capturing Materials: A Review of Their Recent Modification Approaches and Implementation for Industrial Applications
by Himangi Neve, Deven Kadam, Chinmay Sadadekar, Vinayak Wadgaonkar, Siraj Bhatkar and Prasad Wadekar

Preface

Sorption is a ubiquitous natural phenomenon. Although it is a complex mechanism, sorption can be simply explained as a separation process that includes two phases between which some components can be distributed differently. Depending on the nature of the interaction, sorption can be more specifically categorized as either adsorption or absorption (**Figure 1**). Systematic research of sorption processes began in the 18th century, and today numerous theories, mechanisms, and models of sorption processes and behaviors have been developed. Different analytical methods, mathematical models based on sorption theories and mechanisms, and new findings about sorption processes have led to the development of new sorption technologies that contribute to pollution control, environmental protection, and climate change mitigation. The high demand for high-performance sorbents for various applications presents challenges for the research community. The usage of sorption technologies, as well as their integration with other methods, can contribute to improving efficiency, reducing costs, and promoting sustainability.

This book presents research on sorption in three sections: "Modeling of Sorption Processes", "Novel High-Efficiency Sorbents", and "Sorption Processes for Environmental Protection and Pollution Control".

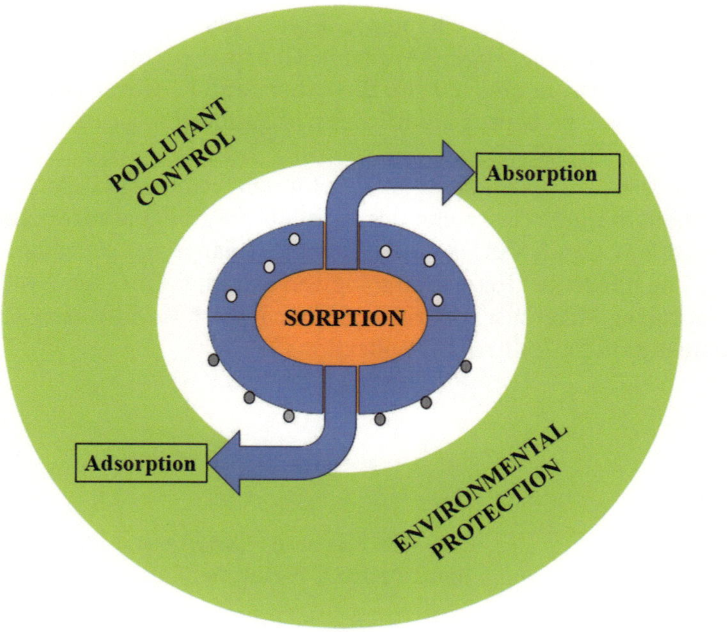

Figure 1.
Sorption mechanisms.

The first section includes four chapters that discuss adsorption and kinetic parameters, isotherm models, and thermodynamic properties of the sorption process.

Chapter 1, "Defluoridation of Contaminated Mine Water by Adsorption on the Fly Ash in the Static and Dynamic Reactor: A Case Study in Morocco", describes the research on the removal of fluoride ions from mine water by adsorption on fly ash from a thermal power plant using both static (batch) and dynamic processes.

Chapter 2, "Sorption of Phenolic Compounds from Woodwaste Leachate by Peat Media", presents a study of the sorption mechanism of phenolic compounds in wood waste leachate by peat as well as the application potential of this natural medium.

Chapter 3, "A Fixed-Bed Column Sorption: Breakthrough Curves Modeling", describes mathematical models of isotherms in the fixed-bed column for understanding the adsorption system, evaluation criteria based on statistical criteria, thermodynamic parameters for evaluating the orientation and feasibility of the adsorptive reaction, as well as the possibility of application in scale-up design based on laboratory analysis of parameters.

Chapter 4, "Thermodynamic Properties of Moisture Sorption and Glass Transition of Coconut (*Cocos Nucifera* L.) Powder Fortified with Physiologically Active Components", explores adsorption in CP+FAC, which can be used to determine optimal storage conditions, predict shelf life, and optimize and quantify energy needs during bulking and heat transfer during drying.

Research into new adsorbents for technological applications has intensified in the last decade. The key factors are efficiency, cost-effectiveness, and environmental acceptability. Section 2 includes Chapter 5, "Bottle Gourd (*Lagenaria vulgaris*) Shell as a Natural, Biodegradable, Highly Available, Cheap, Agricultural by-Product, Miscellaneous Biomass, Ion Exchanger, Biosorbent and Fertilizer", discusses one such effective new cationic sorbent, a modification of lignocellulosic biomass called LVS.

The third section discusses the significance of green technology in sorption processes, the development of innovative technologies and biodegradable adsorbents, the materials used for CO_2 adsorption, and their performance and efficiency. It includes Chapter 6, "The Impact of Green Technology on Sorption Processes", and Chapter 7, "Carbon Capturing Materials: A Review of Their Recent Modification Approaches and Implementation for Industrial Applications".

Dr. Karmen Margeta and Dr. Anamarija Farkaš
Institute for Development and International Relations,
Zagreb, Croatia

Section 1

Modeling of Sorption Processes

Chapter 1

Defluoridation of Contaminated Mine Water by Adsorption on the Fly Ash in the Static and Dynamic Reactor: A Case Study in Morocco

Moufti Ahmed, Taoufik Mohamed, Amri Afaf, Mabrouki Jamal, Mountadar Mohammed and Garmes Hocine

Abstract

In this study, we are interested in the elimination of fluoride ions in a batch and dynamic reactor by adsorption on fly ash from the Jorf-Lasfar thermal power plant in El Jadida, Morocco. Indeed, the results obtained showed that the fly ash made it possible to have treated water meeting international standards for fluorides and certain metals such as Pb and Fe; they will undergo a combined treatment. The elimination efficiency increases with the mass of the adsorbent and the stirring speed, with the pH values in a basic medium (7 < pH < 9.5), and this is for a treatment at room temperature. Additionally, it was concluded that fly ash can be used for fluoride removal in waters with high fluoride concentrations. The results obtained in a dynamic reactor have shown that a fluoride elimination percentage of 97% can be reached under the optimal conditions of our experiment. According to its encouraging results by the simplicity proper to the work, the reasonable cost of the works and the equipment and the good efficiency of elimination, it can be foreseen that the fly ash can be used as a filterable adsorbent in the field of water defluoridation.

Keywords: defluoridation, dynamic and static reactor, fly ash, adsorption, pollution

1. Introduction

Adsorption is one of the most widely used techniques. Activated charcoal, the more commonly used, is expensive, which limits its use in countries such as Morocco. Therefore, research has focused on low-cost treatment, while using different adsorbents (clays, industrial waste, etc.) [1]. The advantage of fluoride removal by adsorption on these different materials, compared to other physicochemical techniques in this genre, is that it retains the polluting material without degrading it, thus avoiding the formation of by-products that may be toxic [2, 3].

Thus, several methods are used for the defluorination of water contaminated by fluorides, the main techniques used are precipitation [4, 5], coagulation-floculation

[6–8], electrocoagulation [9, 10], membrane processes [11], adsorption [12–14], and other techniques [15–20].

In phosphate-producing countries (United States, Commonwealth of Independent States, North Africa, and West Africa), the presence of phosphate ores (hydroxy-apatite) is often associated with that of fluorine ores (fluoro-apatite). In these areas, water from confined aquifers may have excessive fluoride levels. The standards of potability in arid climates are 0.8 mg/l (at 25°C) [21–23]. In Morocco, some phosphate regions have many villages where the concentration of fluoride in groundwater exceeds the values recommended by the World Health Organization (WHO) $\gamma(F^-) > 0.8$ mg/l, under continental and semi-continental climate [24–26]. The presence of excess fluoride ions in drinking water is the cause of serious poisoning. Like any trace element, the fluoride ion is necessary and beneficial for the body at low levels (prophylactic effect) but as soon as its concentration is too high, it becomes toxic and leads to dental and bone fluorosis. Fluoride poisoning is serious in some regions of Morocco and represents a real public health problem [27, 28]. Since then, research has been directed to finding an economically feasible method of fluoride removal for regions that have a high concentration of fluoride in their water supplies [29–33].

The aim of this study is to present a method of defluoridation of black phosphate mine water at the city of Youssoufia in Morocco by adsorption on the fly ash in static and dynamic regime.

2. Analytical methods

2.1 Water characterization

In Morocco, the phosphate areas have a geographical area extending from Khouribga, Oued Zem, and Tadla to the northeast, Settat to the center, and Ben Guerir and Youssoufia to the west (**Figure 1**) [34]. The choice of the Youssoufia area is justified on the one hand by the fact that the latter contains significant concentrations of fluoride in the order of 3 mg/l on average in mine water, and on the other hand, that all the studies have already been done on other regions [35].

The mine water of black phosphates corresponds to groundwater pumped from several superimposed aquifers drowning the different layers of phosphates in the southern zone of the deposits at Youssoufia [36, 37]. This water, which could not be considered as wastewater, is discharged at a very high flow rate of about 8000 m^3/day, up to 35,000 m^3/day. All water samples are stored at 4°C in plastic bottles and analyzed in the laboratory within 24 h according to French standard methods (AFNOR) [36, 38]. Mine water was taken from the drainage channel. The techniques and methods used are summarized in **Table 1**.

2.2 Chemical analysis of fly ash

The chemical analysis (main constituents) of the ash was carried out using a Siemens type SRX 3000 X-ray fluorescence spectrometer, equipped with a tube with Rhodium anode. In addition to this, fly ash is also characterized by thermogravimetric and differential analyses (TGA-DTA), X-ray diffraction measurements (DRX), infrared spectrum (IR) Brunauer, Emmet and Teller (BET) surface measurement and scanning electron microscopy (SEM) [37].

Figure 1.
Map of the major phosphate basins in Morocco.

Parameters and unit	Standards	Serial number NFT	Method principle/device reference
pH (pH unit)	Physical settings		Electrometric/pH meter type WTW 522
EC (ms/cm)			HACH conductivity meter, model 44600
F^- (mg/l)	[38]	90-004	Potentiometric, TISAB* Solution/ pH Orion model SA 520
Cl^- (mg/l)	[38]	90-014	Mohr volumetric method
NO_3^- (mg/l)	[38]	90-012	Sodium salicylate method
SO_4^{2-} (mg/l)	[38]	90-009	Nephelometry method
K^+ (µg/mL) and Na^+ (µg/mL)	[36, 38]	90-019	Atomic absorption spectrometry (ICP-AES)**
Ca^{2+} (mg/l)	[38]	90-016	Volumetric determination by EDTA***
Mg^{2+} (mg/l)	[38]		Difference between calcium and hardness
HCO_3^- (mg/l)	[38]	90-036	Alcalimetric

**TISAB: total ionic strength adjustment buffer.*
***ICP-AES: inductively coupled plasma-atomic emission spectrometry.*
****EDTA: ethylene diamine tetra acetic.*

Table 1.
Techniques and methods used for water characterization.

2.3 Adsorption of fluoride ions, mine water, batch and dynamic

2.3.1 Batch adsorption

Defluoridation tests were carried out on crude ash by introducing 10 g/l of each component into a beaker containing the water to be treated (pHi = 7.86 and $\gamma(F^-)$ = 2.50 mg/l). All of this is subjected to mechanical agitation (V = 300 rpm) by a Jart-Test bench, model JF/6), and the tests were carried out at ambient temperature.

2.3.2 Continuous adsorption (column)

After encouraging results obtained after defluoridation of mine water on a small column in continuous operation, we found it useful to use a larger column than the one used in the first experience. The new column used for the defluoridation of Youssoufia mine water is a glass column 48 cm high and 3.3 cm in diameter. The tests were undertaken on a mixture of fly ash ($m1$ = 70 g) and medium fine sand ($m2$ = 70 g), the tests were carried out at room temperature. The ash used is washed several times until the concentration of fluoride ions released into the wash water is close to 0 mg/l, the average fine sand has a particle size between 160 and 250 μm. The device used is schematized in **Figure 2**. Sand was used to avoid clogging problems. The concentration of fluoride in the effluent was determined every 24 h according to the method described above (**Table 1**).

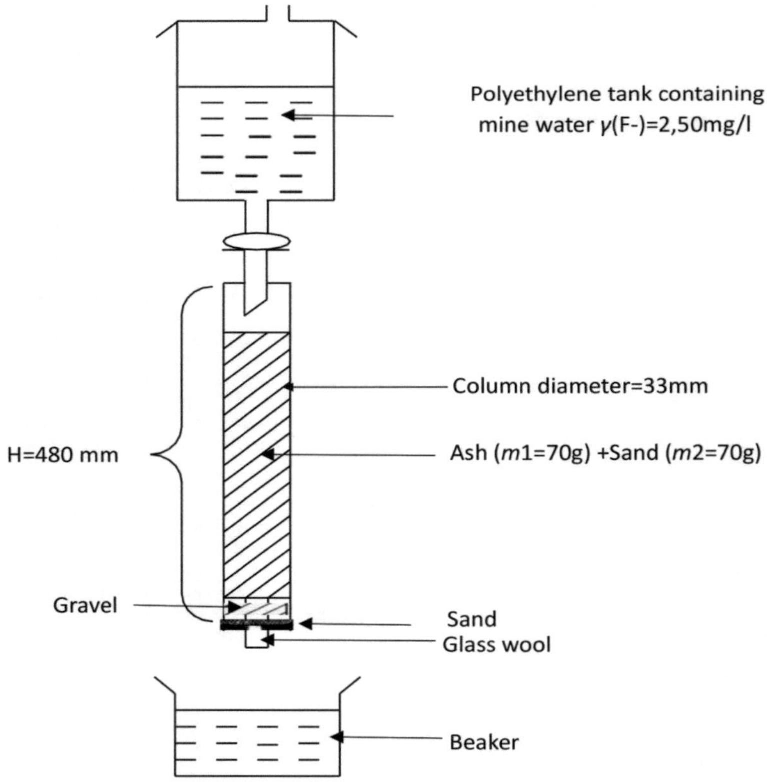

Figure 2.
The pilot used for mine water defluoridation.

3. Results and discussion

3.1 Characterization of Youssoufia mine water

Youssoufia's mine waters are natural waters characterized by a variable chemical composition. **Table 2** shows the average physic-chemical analyses of selected parameters. Indeed, these waters have fluoride levels very concentrated compared to standards [38, 39].

3.2 Chemical composition of ash

In the city of El Jadida in Morocco, located 115 km from the city Youssoufia, we have large quantities of fly ash from the Jorf-Lasfar thermal power plant. **Table 3** shows the chemical composition of fly ash from the Jorf-Lasfar thermal power plant in El Jadida. Note that silica SiO_2 and alumina Al_2O_3 exist in very large quantities. This type of ash rich in silica and alumina is called silicoaluminous ash, and it belongs to class F according to American Standard Methods (ASTM) files [40–42].

3.3 Defluoridation of mine water

3.3.1 Batch reactor (static regime)

3.3.1.1 Kinetics of removal of F^- by the different constituents of the ash

Defluoridation tests were carried out on crude ash, $CaCO_3$, Silica, and Fe_2O_3 by introducing 10 g/l of each component into a tray containing the water to be treated (pHi = 7.86 and $\gamma(F^-)$ = 2.50 mg/l). All of this is subjected to mechanical agitation (V = 300 rpm) by a Jart-Test bench, model JF/6, the tests were carried out at room temperature. The results concerning the kinetics of fluoride removal on the

Parameters	**pH**	**EC* (ms/cm)**	**$\gamma(F^-)$** (mg/l)**	**Cl^- (mg/l)**	**SO_4^{2-} (mg/l)**	**NO_3^- (mg/l)**
Average values	7.67–8.10	1.21–1.34	2.4–4.4	173–197	140–179	9–32
Parameters		HCO_3^- (mg/l)	Na^+ (mg/l)	K^+ (mg/l)	Ca^{2+} (mg/l)	Mg^{2+} (mg/l)
Average values		281–305	58–87	3–12	60–100	55–88

**EC: electrical conductivity.*
***$\gamma(F^-)$: mass concentration of fluoride.*

Table 2.
Average chemical composition of mine water analysis.

Oxide	**SiO_2**	**Al_2O_3**	**TiO_2**	**Fe_2O_3**	**Cao**	**Na_2O**	**K_2O**	**MgO**	**SO_3**	**P_2O_5**
% mass	50	25	1.5	6	6	0.34	1.5	1	1.5	0.17

Table 3.
Chemical composition of fly ash from the Jorf-Lasfar thermal power plant.

different elements are grouped in **Figure 3**. It appears that crude ash and calcium carbonates are the most suitable carriers for defluoridation. Balance is reached after 6 h.

3.3.1.2 Effect of ash washing with HCl on defluoridation

To see the influence of ash washing on defluoridation, we took a well-defined mass of the crude ash, then it was washed with a solution of HCl (1M) while stirring for 3 h. After filtration, the remaining ash was washed with distilled water until the pH was neutral, finally, the remaining mass was placed in the oven at 120°C for 1 day. The defluoridation tests were carried out under the same conditions as before. **Figure 4** shows that ash washed with HCl (1M) does not remove the majority of F^-. This confirms for a second time the participation of the various constituents of the crude ash, and particularly the calcium carbonates.

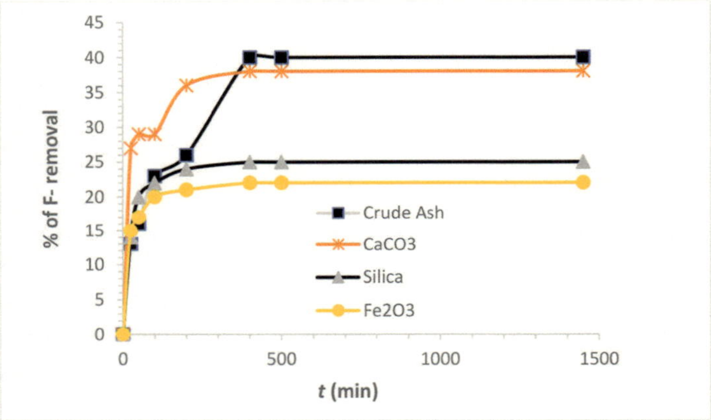

Figure 3.
The kinetics of F^- removal by the different components of the ash.

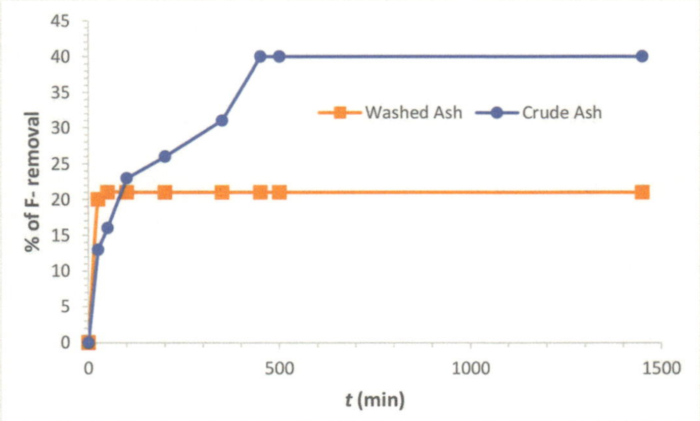

Figure 4.
Kinetics of removal of F^- by crude ash and washed ash.

3.3.1.3 Determination of optimal adsorption conditions

3.3.1.3.1 Influence of stirring rate on adsorption of F^- by crude ash

The influence of stirring rate on the removal of fluorides from mine water by crude fly ash was studied using a constant ash mass (m = 10 g/l) at a constant temperature (T = 22± 3°C). The mixture is subjected to mechanical stirring by a Jart-Test bench for 6 h at different speeds. The results are shown in **Figure 5**.

Figure 5 shows that when the agitation is greater, the percentage of defluoridation is better. This is due to a good dispersion of the ash in the solution.

3.3.1.3.2 Influence of solution pH on removal of F^- from crude ash

To estimate the influence of pH on the F^- removal process by crude ash, a series of experiments was carried out under the following conditions:

$$\gamma(F^-) = 2.50\,mg/l,\ m = 10\,g/l,\ V = 300\,rpm,\ T = 22 \pm 3°C\ \text{and}\ t = 6\,h.$$

The pH of the solution is adjusted with HCl (3M), HCl (6M), and NaOH (6M) solutions. The results are shown in **Figure 6**.

According to **Figure 6**, alkaline pH waters are the easiest to treat by ash, and this is the case for Youssoufia mine water (pH > 7.92). Indeed, starting from a given initial pH, it increases during adsorption. This is due to the solubilization of CO_3^{2-} ions of crude ash. The same results were observed by Moufti et al. [43], which is when adsorption is of the same pollutant on cellulose, alkaline waters are the easiest to treat.

3.3.1.3.3 Effect of temperature on removal of F^- by crude ash

To control the influence of temperature, measurements were made after 6 h of stirring at 15°C, 22°C, 30°C, and 40°C using a crude ash mass equal to 10 g/l, and

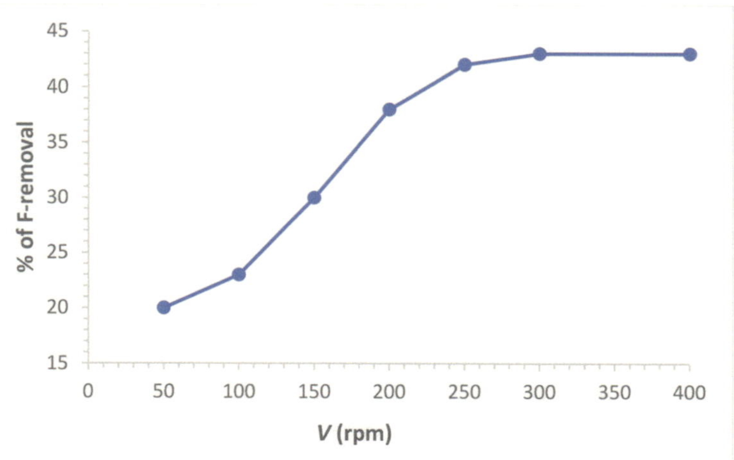

Figure 5.
Influence of stirring rate on the removal of F^- by crude ash ($\gamma(F^-)$ = 2.50 mg/l, pH = 7.86 and t = 6 h).

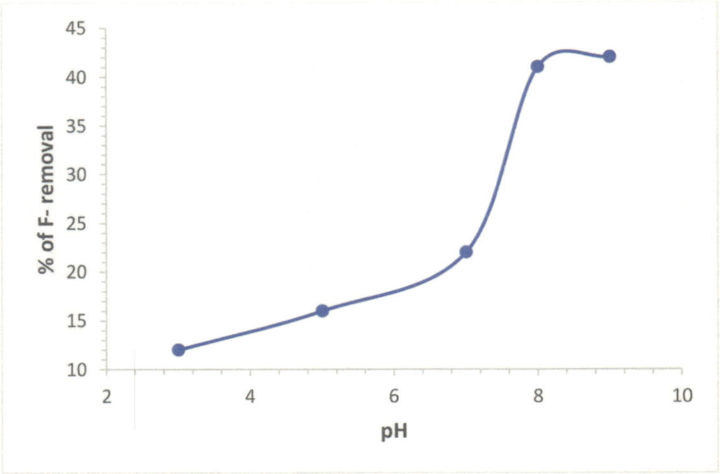

Figure 6.
The influence of the initial pH on the removal of F^- by crude ash: ($\gamma(F^-)$ = 2.50 mg/l, m = 10 g/l, t = 6 h and V = 300 rpm).

the stirring rate is equal to 300 rpm. **Figure 7** shows that the percentage of fluoride removal for a temperature T = 23 C° (ambient temperature) is significant compared to other temperatures.

3.3.1.3.4 Influence of ash mass on the defluoridation process

The influence of the mass of crude ash on fluoride removal was studied, working with masses ranging from 1 g/l to 10 g/l; the stirring time is equal to 6 h, V = 300 rpm, pH = 7.86, $\gamma(F^-)$ = 2.50 mg/l, and t = 22°C. The results are shown in **Figure 8**.

The curve in **Figure 8** shows that the greater the mass of the adsorbent, the higher the fluoride removal rate. The same behavior was observed by Garmes [35].

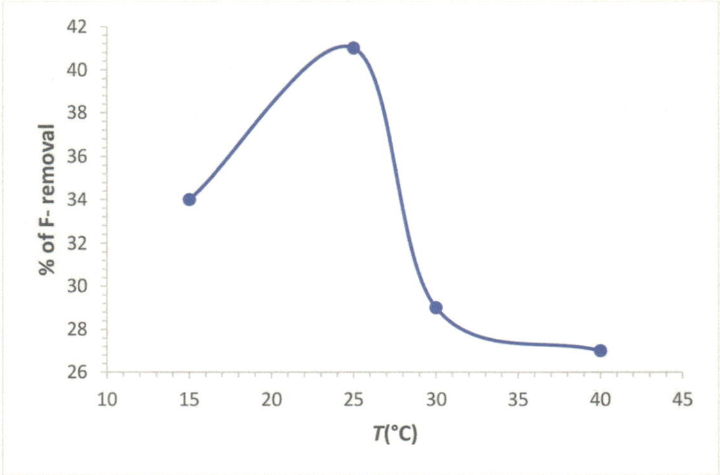

Figure 7.
The influence of temperature on the removal of F^- by crude ash ($\gamma(F^-)$ = 2.50 mg/l, m = 10 g/l, pH = 7.86, t = 6 h and V = 300 rpm).

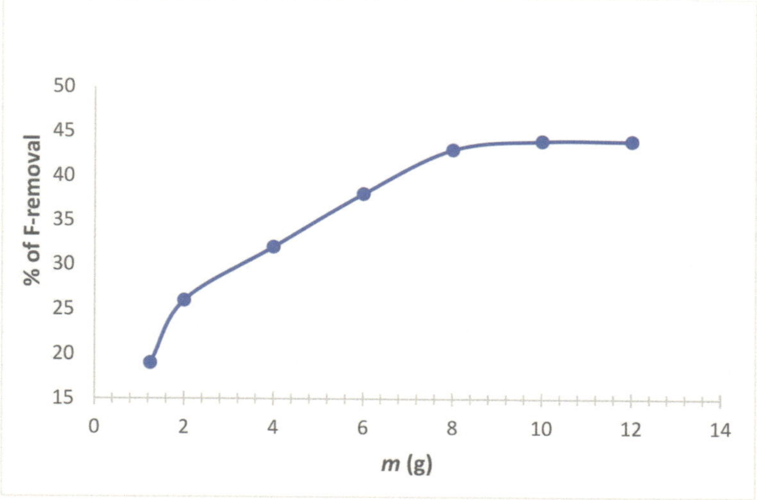

Figure 8.
Effect of ash mass on fluoride removal from mine water by crude ash.

3.3.1.3.5 Influence of the initial concentration of F^- on the defluoridation process

The influence of the initial concentration of F^- on the removal of this pollutant on fly ash was studied by working with different concentrations ranging from $\gamma 0$ = 2 mg/l to 6 mg/l under the same conditions m = 10 g/l, t = 6 h, T = 22°C, and V = 300 rpm. The results are shown in **Figure 9**.

Figure 9 shows that concentration has a great influence on the elimination of fluorides, in fact, the more we increase the concentration, the better percentage of elimination is obtained. Therefore, it seems that crude ash is more suitable for fluoride-laden solutions.

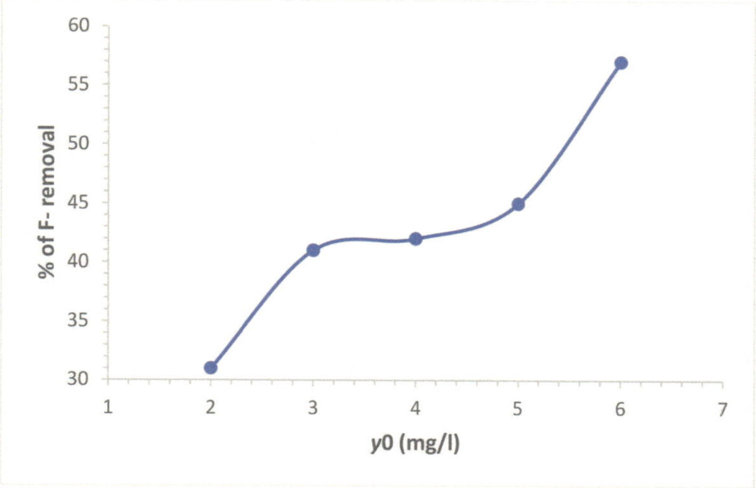

Figure 9.
Effect of initial fluoride concentration on fluoridation defluoridation.

3.3.1.3.6 Langmuir and Freundlich isotherms

In **Figure 10**, we have shown the Langmuir and Freundlich isotherms relating to the adsorption of F^- on fly ash. Linear regression determined the parameters *Qo, b, n,* and *K* (Langmuir and Freundlich constants). The figures obtained show a good correlation, which indicates that these isotherms obey the Langmuir and Freundlich models. The Langmuir and Freundlich constants and correlation coefficients are obtained from Formulas 1 and 2 respectively for cellulose and fly ash. The results obtained are presented in **Table 4**. From these results, it is obvious to conclude that fly ash is more effective than cellulose with a maximum removal capacity equal to 2.38 mg/g and that of cellulose is about 1.18 mg/g according to the Langmuir model. Similarly, for the Freundlich model, the maximum capacity (Qm) obtained by cellulose

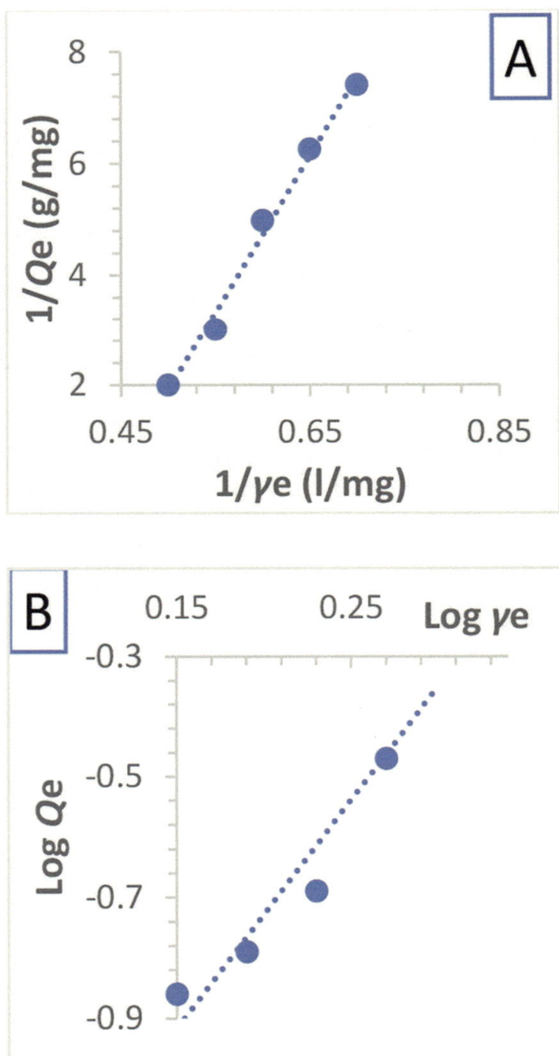

Figure 10.
Linear transformation of the Langmuir (A) and Freundlich (B) isotherms for the couple (F^-/ash) (t = 6 h, pH = 7.86, γo = 2.50 mg/l and V = 300 rpm). Qe: equilibrium adsorption capacity, γe: equilibrium concentration.

Isotherm	Langmuir			Freundlich			
Parameter	Qm (mg/g)	*b* (l/mg)	R^2	*K* (l/g)	*n*	Qm (mg/g)	R^2
Ashes	2.38	14	0.99	2.40	4.03	2.60	0.98

Table 4.
Langmuir and Freundlich isotherm parameters for fluoride adsorption on fly ash.

(Qm = 0.14 mg/g) is lower than that obtained for coal fly ash (Qm = 2.60 mg g) [1]. Therefore, we can conclude that our adsorption is physisorption type and that most of the ash constituents are involved in this phenomenon.

$$\text{For Langmuir (Plot A): the equation is } Y = 0.42332 + 0.037008X \text{ and } R = 0.99297, \quad (1)$$

$$\text{for Freundlich (Plot B): the equation is } Y = 0.38674 + 0.25499X \text{ and } R = 0.98234. \quad (2)$$

3.3.1.3.7 Characterization of treated water before and after ash defluoridation

To understand the quality of water treated (mine water) by ash, we carried out analyses of certain heavy metals by atomic absorption spectrometry. The results of these analyses are collected in **Table 5**. The results of trace metal analyses, carried out by atomic absorption, show that there is a slight increase for some elements, but they remain comparable with those recommended by the National Office of Drinking Water (NODW) in Morocco, except for lead (Pb) and iron (Fe) that exceed the limit even before treatment.

3.3.1.4 Fluoride-concentrated water defluoridation test by the ashes

Because of the advantages of fly ash mentioned earlier as an inexpensive adsorbent, we thought it would be interesting to study its effectiveness in removing fluorides from waters with high concentrations of F^-. We carried out a series of experiments under the following conditions: t = 24 h, V = 300 rpm, and T = 22 °C, but this time with synthetic waters with different initial concentrations of F^- ranging from 2.5 to 50 mg/l and different masses of ash. The first tests did not give good results (**Table 6**).

According to **Figure 11** and **Table 6**, it is shown that if the mass of the adsorbent is doubled, the percentage of removal of F^- is important and can reach 97% for $\gamma 0$ = 58 mg/l for a mass of ash equal to 200 g/l.

Element	Zn (mg/l)	Cu (mg/l)	Pb (mg/l)	Fe (mg/l)	Co (mg/l)
Water before treatment	0.173	0.0013	0.336	0.910	0.310
Water after treatment	0.199	0.1666	0.475	1250	0.207
Limit acceptable to NODW (mg/l)	5	1	0.05	0.3	—

NOWD, National Office of Drinking Water in Morocco.

Table 5.
Metal analyses of mine water treated with defluoridation before and after treatment.

Initial concentration of F^- (mg/l)	Mass of ash (g/l)	% F^- removal
2.50	10	42
30	20	70
45	40	86
52	80	94
58	200	97

Table 6.
Results of defluoridation of synthetic water-high concentered in fluoride by coal fly ash.

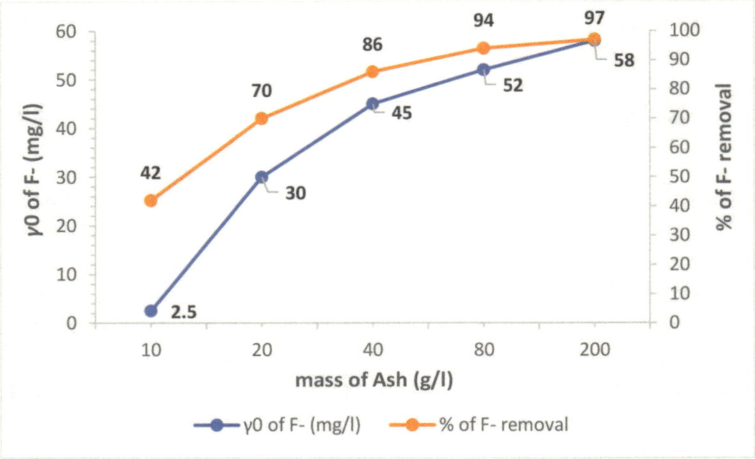

Figure 11.
Correlation between the mass of the adsorbent, the initial concentration of the adsorbate, and the percentage of F^- removal in high-concentration waters.

In conclusion of this part in this study (adsorption of fluoride by ash in static regime), we can say that the results of the adsorption of fluorides from the groundwater of Exhaure (Youssoufia-Morocco), in our experimental conditions, are better compared to other studies carried out in our lab with other adsorbents (cellulose, chitosan) [31, 35, 44] and with other studies on this same waste [45–47]. This treatment has allowed to have a treated water that meets the international and national standards of Morocco.

3.3.2 In continuous reactor (on column)

After studying the mechanisms of static adsorption, which has made it possible to optimize the operating conditions for the elimination of fluoride ions on fly ash, the interest of our study has focused on dynamic sorption performance to find an industrial-scale approach to the use of fly ash. Adsorption experiments of F^- ions soluble in mine water were undertaken at 25°C in a column (**Figure 12**) filled with a well-defined mass of crude fly ash. The process was as follows: 250 ml of the solution was poured the first time on the column to moisten the ash. Then a 250 ml part of the solution was placed in the upper tank, and the flow rate was adjusted between 4.6, 7.7, and 15.4 ml/h. The concentration of fluoride in the effluent was

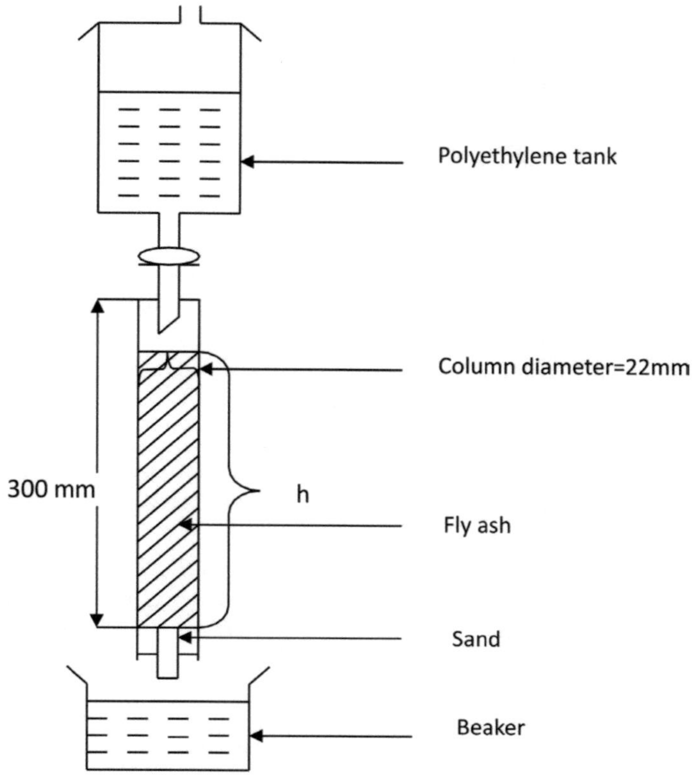

Figure 12.
Scheme of the assembly used for dynamic defluoridation of mine water.

potentially determined every 24 h. The initial concentration of F^- ($\gamma(F^-)$) is equal to 2.50 mg/l and pH = 7.86.

The results are summarized in **Table 7**. The level of fluoride in the effluent gradually decreased to 0 mg/L, when the flow rate decreased from 15.4 to 5.2 to 4.8 ml/h after 96–120 h (**Table 7**). The same results were found by Piekos et al. [40] in a study conducted on the removal of fluorides in synthetic waters with the following concentrations: 1, 5, 10, 20, 50, and 100 mg/l in a column (d = 45 mm and l = 400 mm) filled with 450 g of ash. The flow rate of the column is equal to 2 ml/h.

After these encouraging results, we found it useful to use a larger column than the one used the first time. The new column used for the defluoridation of Youssoufia mine water ($\gamma(F^-)$ = 2.50 mg/l) is a glass column 48 cm high and 3.3 cm internal diameter. The tests were undertaken on a mixture of fly ash ($m1$ = 70 g) and medium fine sand ($m2$ = 70 g), and the tests were carried out at room temperature. The ash used is

t (h)	0	24	48	72	96	120	Column parameters
$\gamma(F^-)$, mg/l	2.5	1.7	1.1	0.5	0.16	0	Flow rate = 15.4 ml/h
	2.5	1.2	0.7	0.4	0	0	Flow rate = 7.7 ml/h
	2.5	0.9	0.45	0.1	0	0	Flow rate = 4.8 ml/h

Table 7.
Results of mine water defluoridation in dynamic reactor for different ash masses.

washed several times until the concentration of F^- ions released into the wash water is close to 0 mg/l, and the average fine sand has a particle size between 160 and 250 μm. The device used is schematized in **Figure 12**.

3.3.2.1 Void ratio and residence time of column operation

The mine water used in this part of the study has an initial concentration equal to 2.50 mg/l and a pH = 7.86. Before fixing the feed rate, it was necessary to determine the effective porosity corresponding to the column. The effective porosity or vacuum index is equal to the volume of water recovered on the total volume of the column [35, 48, 49]. The characteristics of the column are grouped in **Table 8**.

3.3.2.2 Defluoridation of water in dynamic regime (column)

The results of the column (ash + sand) obtained are schematized in **Figure 13**. The level of fluoride in the effluent decreased from 2.50 mg/l to 0.34 mg/l for a time of 15 h to reach 0.12 mg/l after 24 h.

Fluoride levels in the effluent decreased from 2.50 mg/l to 0.34 mg/l for 15 h to 0.12 mg/l after 24 h, this allowed to have water treated according to standards with less time and using industrial waste stored in nature. These results are very

Void ratio (e)	Dead volume (ml)	Residence time (h)*	Flow rate (ml/h)
0.32	62	3.15	7.5

**h: hours.*

Table 8.
Characteristics of the column used for the defluoridation of our water.

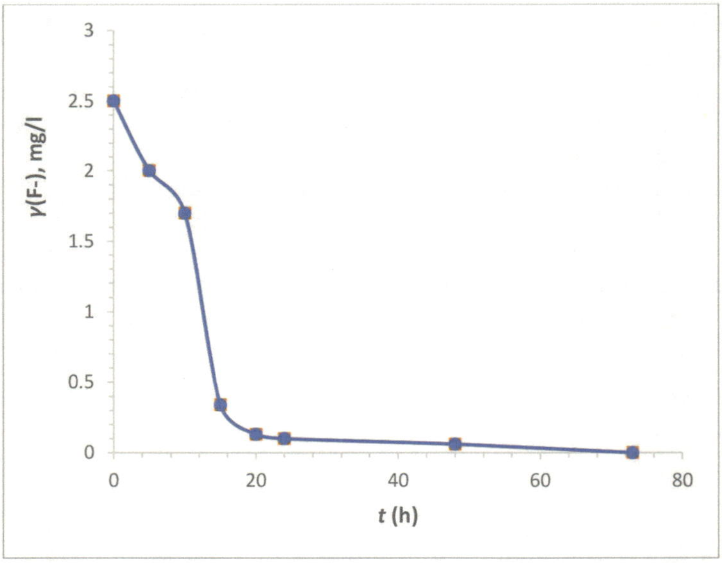

Figure 13.
Results of the defluoridation of Youssoufia mine water in a dynamic regime.

encouraging compared with others, and few studies are carried out on fluoride adsorption by fly ash per column that have been done elsewhere in the world [50].

4. Conclusion

Adsorption tests carried out in batch reactors have shown that cellulose and fly ash from the Jorf-Lasfar thermal power plant in El Jadida, Morocco can be used for the removal of fluorides present in the Youssoufia mine water. Indeed, the adsorption of fluorides on fly ash is a technique that has made it possible to have treated water that meets international standards for fluorides and for certain metals such as Pb and Fe, they will undergo a combined treatment.

The removal yield increases with the mass of the adsorbent and the stirring speed, and this is for treatment at room temperature. The influence of pH showed that the removal rate also increases with the pH values in the basic medium ($7 < pH < 9.5$). In addition, it was concluded that fly ash can be used for fluoride removal in waters with high fluoride concentrations. Langmuir and Freundlich isotherms can be used to describe the equilibrium of F^- adsorption on fly ash. The maximum adsorption capacities obtained using these isotherms are 2.4 mg/g and 2.60 mg/g for fly ash according to the Langmuir and Freundlich models.

The study carried out on the defluorination of ash-column mine water has revealed some useful conclusions for future semi-industrial applications. The results showed that an F^- ion removal efficiency of 97% can be achieved under our experimental conditions. However, it takes approximately 72–120 h to do this.

All the results obtained during this work lead to the conclusion that the fly ash from the Jorf-Lasfar thermal power plant in El Jadida, Morocco can be used for the removal of soluble fluorides in Youssoufia's mine water. As a result, it is possible on the one hand, to recover these waters dropped into nature by the OCP's group and on the other hand to recover and eliminate these ashes stored in the embankments, because their existence in large quantities causes a permanent risk for the receiving environments.

Author details

Moufti Ahmed[1,2]*, Taoufik Mohamed[1], Amri Afaf[3], Mabrouki Jamal[4], Mountadar Mohammed[2] and Garmes Hocine[2]

1 Physical Science Modeling Team, Regional Center for Education and Training Professions, Casablanca-Settat, Morocco

2 Chemistry Department, Faculty of Sciences, Water and Environment Laboratory, University Chouaib Doukkali, El Jadida, Morocco

3 Materials and Processes Department, Laboratory of Science and Engineering, National School of Arts and Crafts, Team of Innovative and Mechanical Manufacturing Processes, ENSAM, University Moulay Ismail, Meknes, Morocco

4 Laboratory of Spectroscopy, Molecular Modelling, Materials, Nanomaterials, Water and Environment, CERN2D, Faculty of Science, Mohammed V University in Rabat, Rabat, Morocco

*Address all correspondence to: ahmed.moufti@outlook.com

References

[1] Annouar S, Moufti A, Mountadar S, Mountadar M, Soufiane A. The influences of the presence of ions counter on the removal capacity fluorides ions by chitosane. Oriental Journal of Chemistry. 2016;**32**(1):399-406

[2] Obijole O, Wilson G, Mudzielwana R, Ndungu P, Samie A, Babatunde A. Hydrothermally treated aluminosilicate clay (HTAC) for remediation of fluoride and pathogens from water: Adsorbent characterization and adsorption modelling. Water Resources and Industry. 2021;**25**:100-144

[3] Kumar P, Gupta P. Defluoridation of domestic wastewater by using activated diatomaceous earth in fixed mattress column adsorption system. Oriental Journal of Chemistry. 2021;**37**:594-601

[4] Dahi E. Contact precipitation for defluoridation of water. In: 22nd WEDC Conference Reaching the Unreached Challenges for 21st Century. New Delhi, India; 1996. pp. 262-265

[5] Consultative Committee of the Kenya Bureau of Standards (KEBS). Excessive Fluoride in Water in Kenya, Nairobi, Kenya; 2010

[6] Hu CY, Lo SL, Kuan WH. Effects of the molar ratio of hydroxide and fluoride to Al(III) on fluoride removal by coagulation and electrocoagulation. Journal of Colloid and Interface Science. 2005;**283**:472-476

[7] Bolto B, Gregory J. Organic polyelectrolytes in water treatment. Water Research. 2007;**41**:2301-2324

[8] Linzhi Z, Sun Y, Cheng H. Research on coagulation/sedimentation process for simulation of fluorine-containing wastewater treatment. Applied Mechanics and Materials. 2013;**361-363**:755-759

[9] Khan SU, Asif M, Alam F, Khan NA, Farooqi IH. Optimizing fluoride removal and energy consumption in a batch reactor using electrocoagulation: A smart treatment technology. In: Ahmed S, Abbas S, Zia H, editors. Smart Cities – Opportumities and Challenges. Singapore: Springer; 2020. pp. 767-778

[10] Takdastan A, Tabar SE, Islam A, Bazafkan MH, Naisi AK. The effect of the electrode in fluoride removal from drinking water by electro coagulation process. In: International Conference on Chemical and Biological Sciences; 18-19 March 2015. Dubai, UAE; 2015. pp. 39-44

[11] Mohapatra M, Anand S, Mishra BK, Giles DE, Singh P. Review of fluoride removal from drinking water. Journal of Environmental Management. 2009;**91**:67-77

[12] Sun Z, Park J, Kim D. Synthesis and adsorption properties of Ca-Al layered double hydroxides for the removal of aqueous fluoride. Water, Air, & Soil Pollution. 2017;**228**:23-30

[13] Wambu EWEW, Ambusso WOW, Onindo CO, Muthakia GKGK, Wambu EWEW. Review of fluoride removal from water by adsorption using soil adsorbents—An evaluation of the status. Journal of Water Reuse and Desalination. 2016;**6**:1-29

[14] Yapo NZS, Gouesse B, Briton H, Aw S, Reinert L, Drogui P, et al. Toxic/hazardous substances and environmental engineering bivalve shells (*Corbula trigona*) as a new adsorbent for the defluoridation of groundwater by.

Journal of Environmental Science and Health, Part A. 2021;**56**:694-704

[15] Shen J, Mkongo G, Abbt-Braun G, Ceppi SL, Richards BS, Schäfer AI. Renewable energy powered membrane technology: Fluoride removal in a rural community in northern Tanzania. Separation and Purification Technology. 2015;**149**:349-361

[16] Sahli MA, Annouar A, Tahaikt S, Mountadar M, Soufiane A, Elmidaoui A, et al. Fluoride removal for underground brackish water by adsorption on the natural chitosan and by electrodialysis. Desalination. 2007;**212**:37-45

[17] Diawara CK, Diop SN, Diallo MA, Farcy M, Deratani A, Corporation P, et al. Performance of Nanofiltration (NF) and low pressure reverse osmosis (LPRO) membranes in the removal of fluorine and salinity from brackish drinking water. Journal of Water Resource and Protection. 2011;**3**:912-917

[18] Bejaoui I, Mnif A, Hamrouni B. Performance of reverse osmosis and nanofiltration in the removal of fluoride from model water and metal packaging industrial effluent performance of reverse osmosis and nanofiltration in the removal of fluoride from model water and metal packaging. Industrial, Separation Science and Technology. 2014;**49**:1-11

[19] Grzegorzek M, Majewska-nowak K, Ahmed AE. Science of the total environment removal of fluoride from multicomponent water solutions with the use of monovalent selective ion-exchange membranes. Science of The Total Environment. 2020;**722**:137681

[20] Dubey S, Agarwal M, Gupta AB. Experimental evaluation of sand filtration and ultrafiltration as subsequent treatment of coagulation for fluoride removal. Environmental Progress & Sustainable Energy. 2021;**e13790**:1-11

[21] Moufti A. Physico-chemical study of Youssoufia mine water (Morocco) and their defluoridation by adsorption on cellulose and coal fly ash [doctoral thesis]. El Jadida; Morocco: Chouaib Doukkali University; Faculty of Sciences; 2003

[22] Waugh DT, Potter W, Limeback H, Godfrey M. Risk assessment of fluoride intake from tea in the Republic of Ireland and its implications for public health and water fluoridation. International Journal of Environmental Research and Public Health. 2016;**13**:259-280

[23] Mountadar M, Garmes H, Abouragi M, Yousrani K. Contamination of groundwater and soil by fluorides near phosphate mining plants (MOROCCO). Journal of the European Scientific Association for Water and Health. 2000;**5**(1):61

[24] Levy SM. An update on fluorides and fluorosis. Journal of the Canadian Dental Association. 2003;**69**:286-291

[25] UNICEF. WHO, Progress on sanitation and drinking water. 2010 Update. 2010. Available from: http://www.unwater.org/downloads/JMP_report_2010.pdf

[26] Jacintha TGA, Rawat KS, Mishra A, Singh SK. Hydrogeochemical characterization of groundwater of peninsular Indian region using multivariate statistical techniques. Applied Water Science. 2017;7:3001-3013

[27] Fandi R. Study of the relationship between fluoride levels in water, soil, and human urine in the Darmous regions (Youssoufia) [doctoral thesis]. Pharmacy, Mohammed V University;

Faculty of Medicine and Pharmacy of Rabat; 1994

[28] Yousrani K. Epidemiological investigation on the evolution of dental fluorosis of dental fluorosis in the province of Khouribga [thesis medicine]. Casablanca, Morocco: Hassan II University; 1990

[29] Ayoob S, Gupta K. Fluoride in drinking water: A review on the status and stress effects. Critical Reviews in Environmental Science and Technology. 2006;**36**:433-487

[30] Mountadar M, Garmes H, Moufti A. Defluoridation of groundwater-Youssoufia-Maroc. Annale Journal Chemistry Science Materials. 2001;**26**:S341-S344

[31] Mountadar M, Garmes H, Bouraji M. Defluoridation of fluoride-laden water: The case of the discharge from the phosphate laundromat. Physical Chemistry News (Khouribga, Morocco). 2005;**25**:88-92. In: Moufti A et al., editors. Proceedings of the International Conference on Membranes and Separation Processes (CIMP); 1. Morocco; 1999. pp. 89-94

[32] Moufti A, Mountadar M. Valorization of a fly ash by the defluoridation of groundwater, case of Youssoufia-Morocco. In: 2nd Symposium of the Moroccan Group for Research for the Environment (GMRE) and the XXXII Congress of the French Group of Pesticides (GFP); 29-31 May; Marrakech, Morocco. 2002

[33] Botsa SM, Basavaiah K. Defluoridation in aqueous solution by a composite of reduced graphene oxide decorated with cuprous oxide via nonochemical. Arabian Journal of Chemistry. 2020;**13**:7970-7977

[34] Rempert TH, Martens BP, et al. First record of a tylosaurine mosasaur from the latest cretaceous phosphates of Morocco. Open Journal of Geology. 2022;**12**(11):883-906. DOI: 10.4236/ojg.2022.1211042

[35] Garmes H. L'analyse des interférences naturelles et anthropiques sur le développement de la fluorose (province de Khouribga); Défluoruration par des procédés conventionnelles et membranaires [Thèse de Doctorat d'état]. El Jadida, Morocco: Université Chouaib Doukkali, Faculté des Sciences; 2002

[36] Rodier J. Analysis of Natural Water, Wastewater, Seawater. 8th ed. Paris: Dénod; 1996

[37] Moufti A, Brahmi R, Garmes H, Bensitel M, Mountadar M. Characterization of a flying ash stemming from the combustion of the coal. In: Séminaire International sur la physicochimie des Matériaux (REMCES 9); 30-31 October et 1er November 2002. Agadir, Morocco; 2002

[38] AFNOR. Collection of French Standards: Water Quality. 3rd ed. 1999

[39] Moufti A. Characterization of the Youssoufia-Morocco-mine fluoride-contaminated water and their detrimental effects on human health. In: Environmental Chemistry and Recent Pollution Control Approaches. No. 1. 2019. pp. 3-20. DOI: 10.5772/intechopen.80547

[40] Piekos R, Paslawska S. Fluoride uptake characteristics of fly ash. Fluoride. 1999;**32**(1):14-19

[41] Roy WR, Theiry RG, Sculler RM, Suloway JJ. Coal fly ash: A review of the literature and proposed classification system with emphasis on environmental impact. Environmental Geology. 1981;**96**:1-69

[42] Minoux MA. Fly ash and micro ash: Production processes, physical and

chemical consequences on the micro-ash-cement-water system. Civil Engineering Thesis, Toulouse. 1994;**271**:225

[43] Moufti A, Annouar S, Mountadar S, Mountadar M. The regeneration of the pre used ashes in the elimination fluorides ions from the underground waters. Journal of Materials and Environmental Science (Morocco). 2016;7(6):2069-2073

[44] Mountadar M, Garmes H, Bouraji M. Defluoridation of fluoridated water loaded with fluorides: Case of the discharge of the phosphate laundromat (Khouribga-Morocco). In: Proceedings of the International Symposium on Membranes and Separation Processes (CIMP). Vol. 1. 1999. pp. 89-94

[45] Ye C, Yan B, Ji X, Liao B, Gong R, Pei X, et al. Adsorption of fluoride from aqueous solution by fly ash cenospheres modified with paper mill lime mud: Experimental and modeling. Ecotoxicology and Environmental Safety. 2019;**180**(30):366-367

[46] Obrenovie JVM, Smiljanie S. Application of fly ash for flouride adsorption. Zastita Materijala. 2022;**63**(4):395-403

[47] Ramesh ST, Gandhimathi R, Nidheesh PV, Taywade M. Batch and column operations for the removal of fluoride from aqueous solution using bottom ash. Environmental Research Engineering and Management. 2012;**2**(2):12-20. DOI: 10.5755/j01.erem.60.2.1396

[48] El Krati M. State of the environment of the city of El Jadida-Study of the treatability of liquid textile discharges [doctoral thesis]. Morocco: Chouaib Doukkali University, Faculty of Sciences, El Jadida; 2000

[49] El Fakih N. Study of heterotrophic biological denitrification on natural cellulosic substrate-post treatment and water quality [doctoral thesis]. El Jadida. Morocco: Chouaib Doukkali University, Faculty of Sciences; 2000

[50] Ibrahim NK, Khazaal SH. Fluoride removal from aqueous solutions by adsorption with coal ash. Engineering & Technology Journal. 2010;**28**(18):5771-5781. DOI: 10.30684/etj.28.18.13

Chapter 2

Sorption of Phenolic Compounds from Woodwaste Leachate by Peat Media

Najat Kamal, Rosa Galvez, Gerardo Buelna and Abdelaziz Bacaoui

Abstract

This study contributes to the clarification of sorption mechanism of phenolic compounds in woodwaste leachate by peat, and it is a part of the project which aims to clarify and contribute to determine and evaluate the sorption mechanism part of phenolic compounds in a trickling biofilter. To achieve this objective, mechanisms were studied separately by isolation of each process, and sorption mechanism was followed in the present study by inhibiting the biological process. The kinetic study showed that the maximum sorption capacity was reached between 20 and 24 h at 10°C and between 16 and 20 h at 20°C. However, it is during the first hours that the sorption process is high. The maximum sorption capacity was evaluated at 68.5 mg/kg (57.87% of the initial concentration) for the most polar compounds: 4-nitrophenol, phenol, and 2-chlorophenol and at 35.2 mg/kg of peat for the least polar compounds such as 2,4-dimethylphenol under conditions of pH 4 and at 10°C. The description of sorption results was evaluated by a kinetic and thermodynamic study and modeling by Langmuir and Freundlich isotherm.

Keywords: phenolic compounds, peat moss media, sorption capacity, kinetics, Freundlich and Langmuir models, parameters interaction

1. Introduction

In Quebec, according to the directory of primary wood processing plants, published by the Ministry of Forests, Wildlife and Parks (MFFP) in August 2014, materials derived from wood (sawing, panels and veneers, and energy materials) bring together around 400 establishments, of which nearly a hundred have an annual consumption of more than 100,000 m^3 of wood [1]. Environmental challenges are related to the generated volume of water process and to the leachate from onsite woodpiling. A recent review done in January 2023, replacing 2015 version and confirming this issue with the highlight of the risk of groundwater contamination, atmospheric emissions, and odors, related the presence of lignin compounds [2].

Rainwater that percolates raw materials, process equipment, process or residual materials leachs substances (e.g., phenols, resin acids, formaldehyde) and/or also by direct runoff (e.g., TSS, hydrocarbons oil tankers C10-C50) [2]. Leachate is generated when water percolates through a large mass of solid material [3–5]. Woodwaste leachate contains lignin and phenolic compounds at important concentrations and must be treated before its discharge in the environment [6].

Phenolic compounds are toxic to humans and ecosystem, are persistent, and are endocrine disruptors. These compounds are the subject of several studies and are under investigation in recent years due to the high potential risk and negative impact on the environment and on human health [7–11]. Their toxicity causes several health risks and problems such as asthma, unconsciousness and cerebral pains, hepatic effects [12], and significant fetal and placental toxicity effects [13]. Phenolic compounds have endocrine-mediated effects in mammals and fish [14] and can disturb the endocrine system and affect reproduction, development, and immune functions [10].

These compounds are released continuously in the environment and aquatic system by the increase of industrial activities and daily needs [15]. In the context of the urgent actions to preserve the environment, reduce impacts, optimize resources, and reduce the greenhouse emissions, all efforts must be put together to find solutions and minimize the negative impact.

Several methods, physicochemical and biological processes, have been proposed for the removal of phenolic compounds from different effluent types and by various matrixes. The commonly used methods are membrane processes [16], solvent extraction [17, 18], oxidation [19, 20], coagulation and electrocoagulation [21], biological processes [22–26], and sorption [11, 27–35].

Sorption is an effective technology and has a great potential due to its advantages including economic, simplicity and ease to use, autonomy, and low maintenance requirements. Many media were explored and studied as sorbents to remove phenolic compounds from different matrices. The most effective and popular sorbent is the activated carbon, but due to its high cost, researchers continue to develop and test other innovative and sustainable alternative sorbents.

This study aim to clarify, determine, and evaluate sorption mechanisms of phenolic compounds by peat moss media. To achieve this objective, sorption mechanisms were studied separately by inhibiting biological processes.

The isolated sorption mechanism of phenolic compounds by peat moss was studied in batch tests at two fixed values of pH and temperature.

2. Material and methods

Experiments were performed at small scale using batch tests described by **Figure 1** and carried out by the following five steps:

- Step 1: Characterization of woodwaste leachate and sorbent

Woodwaste leachate from Quebec area sawmill, the liquid matrix and peat moss, the solid matrix constituting the filter media were characterized. For the leachate, pH and physicochemical parameters that can impact the treatment process, in particular COD, MES, and individual phenolic compounds were determined. For peat moss, mainly parameters influencing the sorption process (particle size, specific surface and cation exchange capacity) were evaluated. The media was then dried at 103°C for 48 h to minimize the bacterial activity.

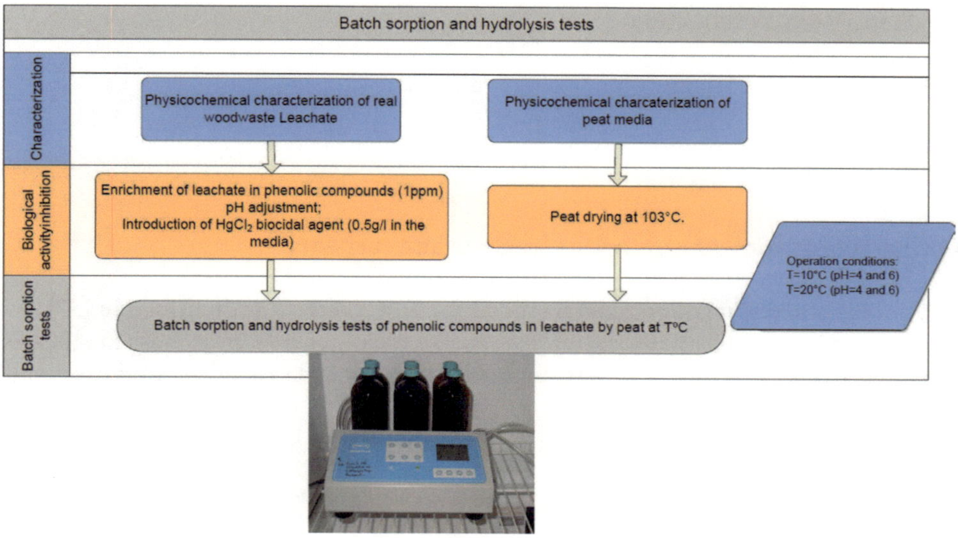

Figure 1.
Batch sorption and hydrolysis tests flow-chart.

- Step 2: Spiked leachate preparation

A preconcentration step with target compounds was carried out to reach the saturation level and follow the variation of their concentration. Woodwaste leachate was spiked to 1 mgL^{-1} of each studied phenolic compounds. The pH was adjusted, and the biological activity was inhibited by adding mercuric chloride solution.

- Step 3: Batch sorption tests and analysis

Amber bottles of 1 L were used for sorption tests. To a volume of 500 mL leachate, different masses of peat were added (m = 0, 5, 10, 15, 20, and 25 g). Tests were followed for 3 days by sampling at different time intervals, each 2 hours during the first 12 hours. Batch experiments were carried out at 10 and 20°C for two pH values: 4 and 6. These operation conditions are the representative conditions of installation operation in Quebec during winter and summer seasons. Collected samples were analyzed to determine residual phenolic compounds concentration.

Samples (40–50 mL) were acidified to pH = 1.8–2, filtered by vacuum on 0.45 μm membrane filter, preconcentrated by the SPE method on HLB Oasis Cartridge (60 mg) conditioned by 2 mL of methanol and 2 mL of water, following the protocol described by [36], eluted by methanol, recovered to 0.2 mL by vacuum evaporation and reconstituted in 1 mL of mobile phase, initial composition for HPLC-UV analysis.

- Step 4: Isotherm modeling

The equilibrium concentrations for each mass of solid studied were determined. Isotherms describing variations of the sorbed concentration versus the equilibrium one were established. The sorption capacity was determined for each studied compound. Freundlich and Langmuir models were tested.

- Step 5: Models validation

Freundlich and Langmuir models were tested.

3. Results and discussion

3.1 Characterization of woodwaste leachate and sorbent

The leachate was characterized by the SPE-HPLC method as described in [37]. **Table 1** summarizes characterization results. On the other side, peat moss media was characterized to determine the pertinent physico-chemical properties, and the results are presented on **Table 2**.

Parameter	In the woodwaste leachate (Quebec area sawmill)
pH	4.50
COD (mg L^{-1})	1350
BOD5 (mg L^{-1})	630
MES (mg L^{-1})	410
PO4 (mg L^{-1})	5.00
Total volatile fatty acids (mg L^{-1})	81
Total phenolic compounds (μg L^{-1})	1979
Phenol (μg L^{-1})	370.65
4-Nitrophenol (μg L^{-1})	268.39
P, m-Cresol (μg L^{-1})	309.43
O- Cresol (μg L^{-1})	133.86
2- Chlorophenol (μg L^{-1})	177.23
2, 4- Dinitrophenol (μg L^{-1})	683.03
2, 4- Dimethylphenol (μg L^{-1})	5.62

Note: Total phenolic compounds were determined by spectrophotometric analysis. Individual phenolic compound values reported in μg L^{-1}.

Table 1.
Physicochemical characterization of real woodwaste leachate.

Parameter	In the peat moss media
pH	3.8–4.6
Phenolic compounds release in water	0
Particles size	0.2–4 mm
Cationic Exchange Capacity (C. E. C)	155 meq/100 g
Nitrogen	0.9% in dried material
Humidity	35–40%

Table 2.
Physicochemical characterization of peat moss media.

The leachate was spiked to 1 mgL^{-1} of each studied phenolic compounds before starting batch sorption tests.

3.2 Batch sorption tests and analysis results

3.2.1 Hydrolysis and peat release tests

Hydrolysis test of phenolic compounds was carried out before adding the sorbent to evaluate the variation of concentration related to the compounds stability in the solution. The observed variation was between 2.86% for phenol and 9.39% for 2,4-dinitrophenol at 10°C for pH 4 and 3.64% for 4-nitrophenol and 13.73% for 2,4-dinitrophenol at 20°C and pH 6. Hydrolysis results for pH 4 and 10°C are shown in **Figure 2**. No release of phenolic compounds was observed under these conditions.

3.2.2 Sorption kinetics

Sorption kinetics for all the selected compounds was followed as function of time. The sorption capacity was determined for each compound in the equilibrium state (**Figures 3–5**). The maximum sorption capacity of phenols on peat was evaluated at 68.5 mg/kg (57.87% of the initial concentration) for the most polar compounds: 4-nitrophenol, phenol, and 2-chlorophenol, and at 35.2 mg/kg of peat for the least polar compounds such as 2,4-dimethylphenol under conditions of pH 4 and at 10°C.

The kinetic study showed that the maximum sorption capacity was reached between 20 and 24 h at 10°C and between 16 and 20 h at 20°C; However, it is during the first hours that sorption process is high.

3.3 Isotherms development and models validation

A simple review of Freundlich and Langmuir theories brings us to the model's equation bellow (**Figure 4**):

3.3.1 Freundlich model

This is an equation established by Freundlich to represent adsorption equilibrium between the solute and the surface of the substrate and is expressed as follows:

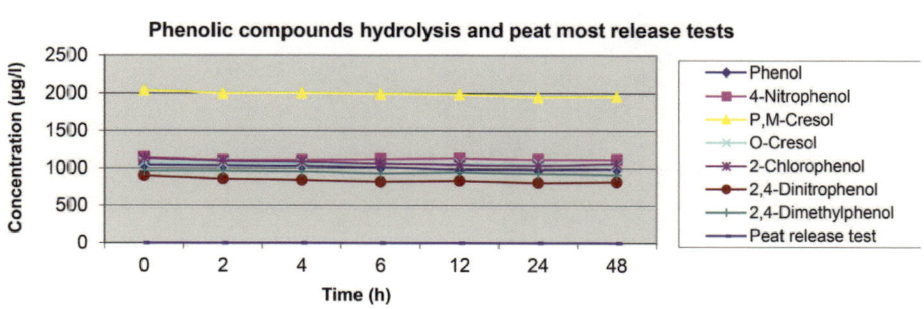

Figure 2.
Phenolic compounds hydrolysis and peat release tests at pH 4 and 10°C.

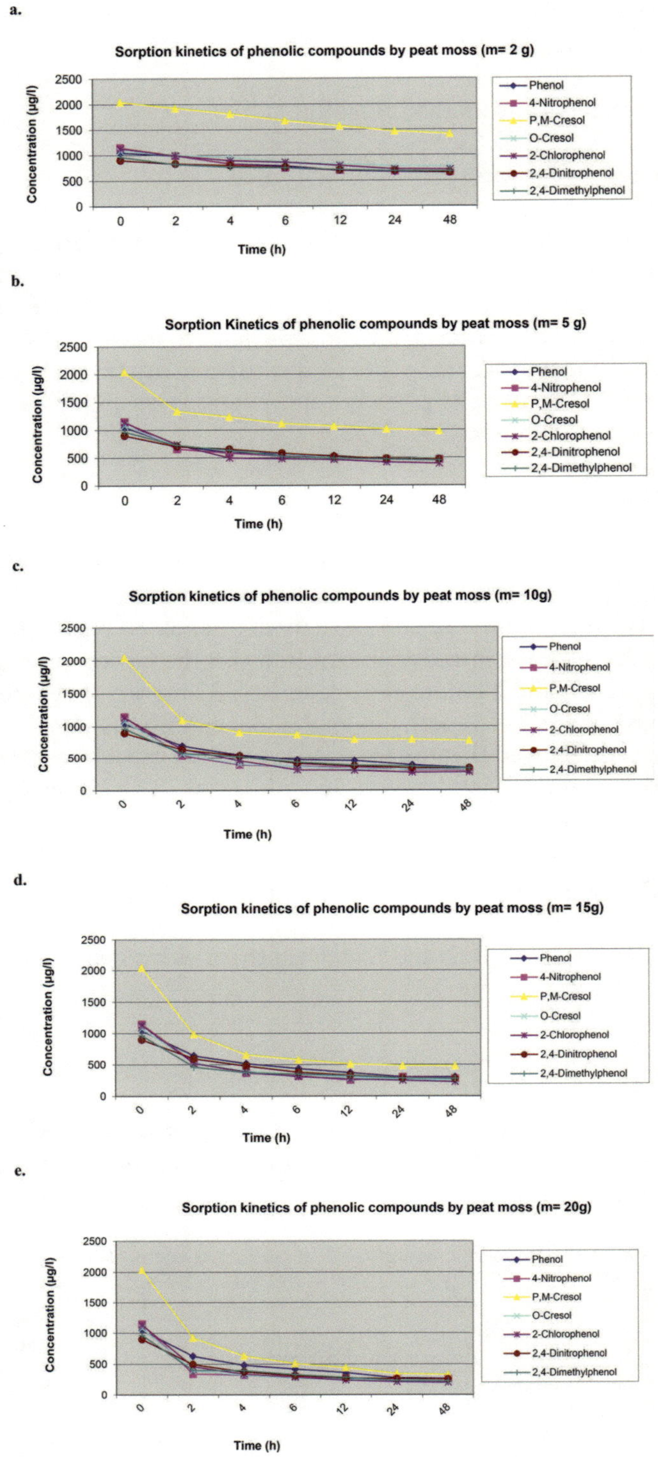

Figure 3.
Sorption kinetics of the studied compounds at pH 4 and 10°C and the effect of the sorbent mass. a. (m = 2 g), b. (m = 5 g), c. (m = 10 g), d. (m = 15 g) and e. (m = 20 g).

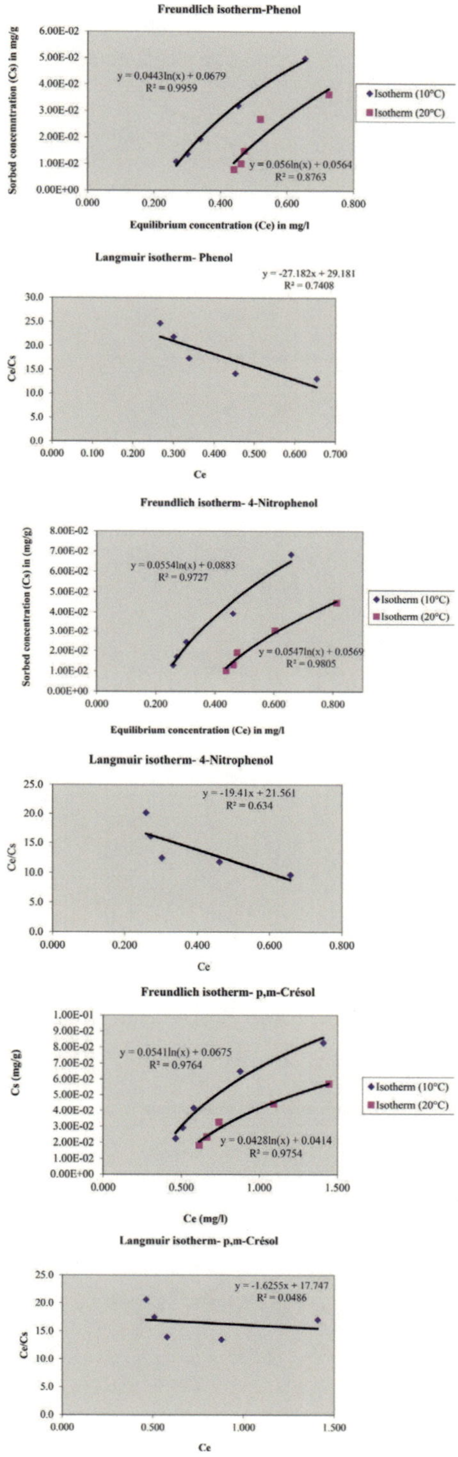

Figure 4.
Freundlich and Langmuir models applied to describe the sorption of the studied phenolic compounds by peat for pH 4 and 10 and 20 C.

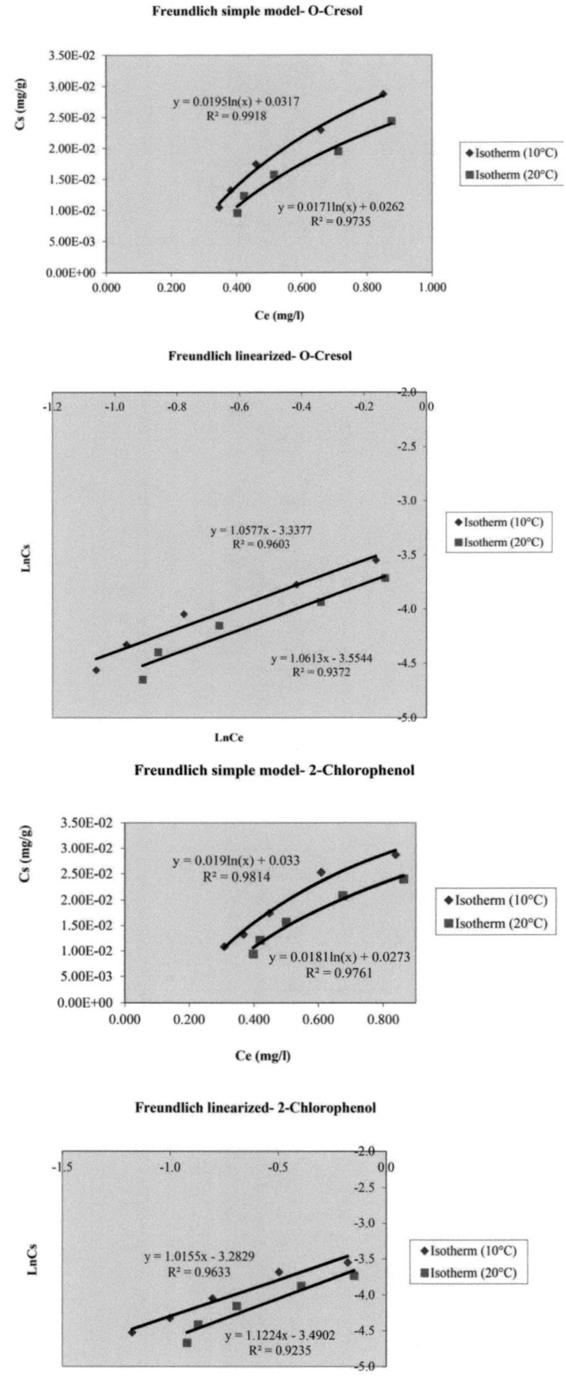

Figure 5.
Examples of correlations obtained with simple and linearized Freundlich model applied to describe the sorption of the studied phenolic compounds by peat for pH 6 and 10 and 20°C.

$$Cs = Q = K_F.Ce^{1/n} \quad (1)$$

Where.

Q or Cs: sorbed quantity of the contaminant (μmol.g^{-1}, μg.g^{-1} ou mg.g^{-1});

K_F and n: Freundlich constant and indicative of the intensity and adsorption capacity.

Ce: residual concentration at the equilibrium (μmol.L^{-1}, μg.L^{-1} ou mg.L^{-1});

The Freundlich linearized form is as follows:

$$LnCs = LnK_F + \frac{1}{n}LnCe \quad (2)$$

The extrapolation of this equation for the initial concentration of the contaminate allow to determine the maximal adsorption capacity (Qm or Csm) in the studied conditions. The isotherm is described by the graph:

$$Cs = \mathrm{f}(Ce) \quad (3)$$

Phenolic compound sorbed amount is expressed as follows:

$$Cs = Q = (C0 - Ce).V/m \quad (4)$$

Where.

C_0: initial concentration (μmol.L^{-1}, μg.L^{-1} ou mg.L^{-1}).

V: volume (L).

m: mass of the sorbent (g).

3.3.2 Langmuir model

Langmuir model is presented as follows:

$$\frac{Ce}{Cs} = \frac{1}{\mathrm{K_L}.Qm} + \frac{Ce}{Qm} \quad (5)$$

Where.

K_L: Langmuir constant (L/mg).

Qm: maximal adsorption capacity (mg/g).

The sorption of all the targeted compounds was described well by Freundlich model with good correlation factors varying from 87 to 99% (**Figure 5**). Low correlation factors were obtained using Langmuir model for all the compounds.

3.4 Statistical analysis and parameters interaction

To evaluate interactions between parameters and their effect on the response, a dependant measure (Y) which is the sorption capacity of each phenolic compound, a design of experiments study was carried out using Design Expert Software.

A preliminary basic statistics review was done using Minitab 21 to explore Scatterplot variation of equilibrium concentration of each compound with the parameter X1 (time). **Figure 6** presents the scatterplot of studied compounds. The results confirm that the sorption capacity process is maximal during the first hours.

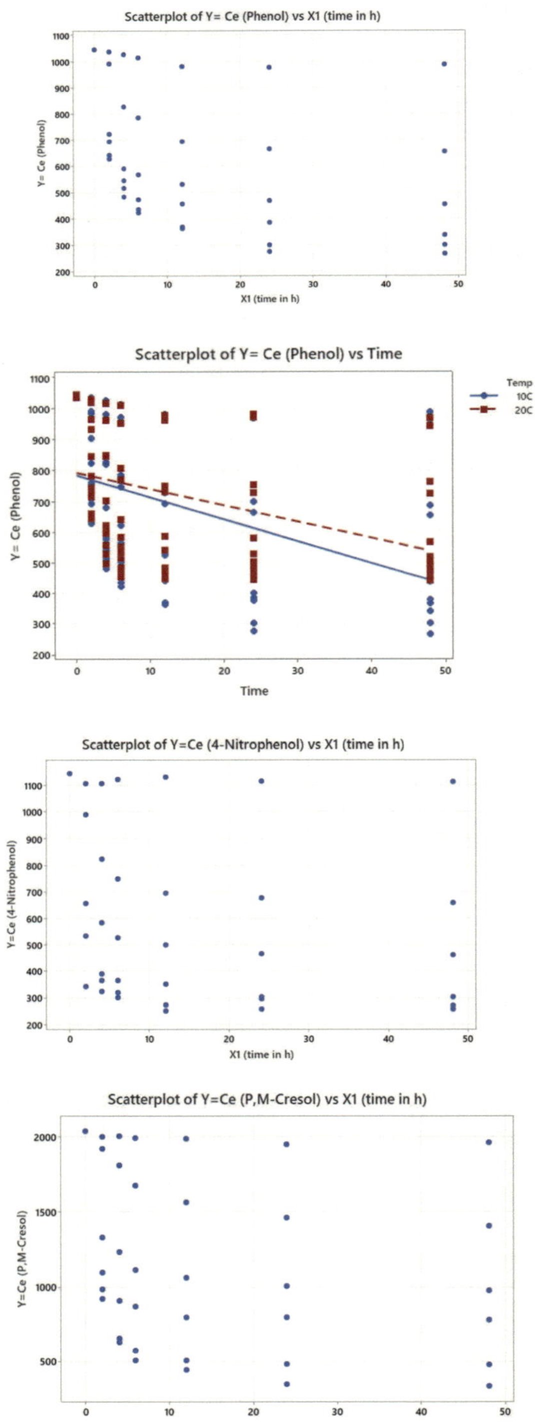

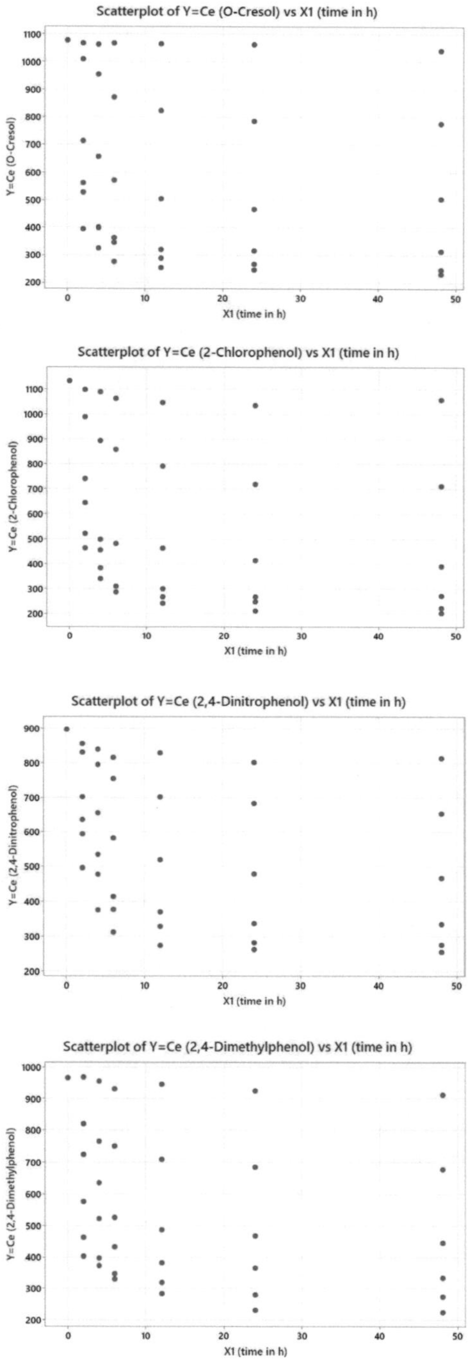

Figure 6.
Scatterplot variation of equilibrium concentration with time for all the studied compounds.

Summary results for phenol are shown in **Figure 7**. Data distribution is not normal (P less than 0.05). The Skewness is good (not very asymmetrical). Results are slightly (shifted) to the left, which can explain a little difference between the mean and the median.

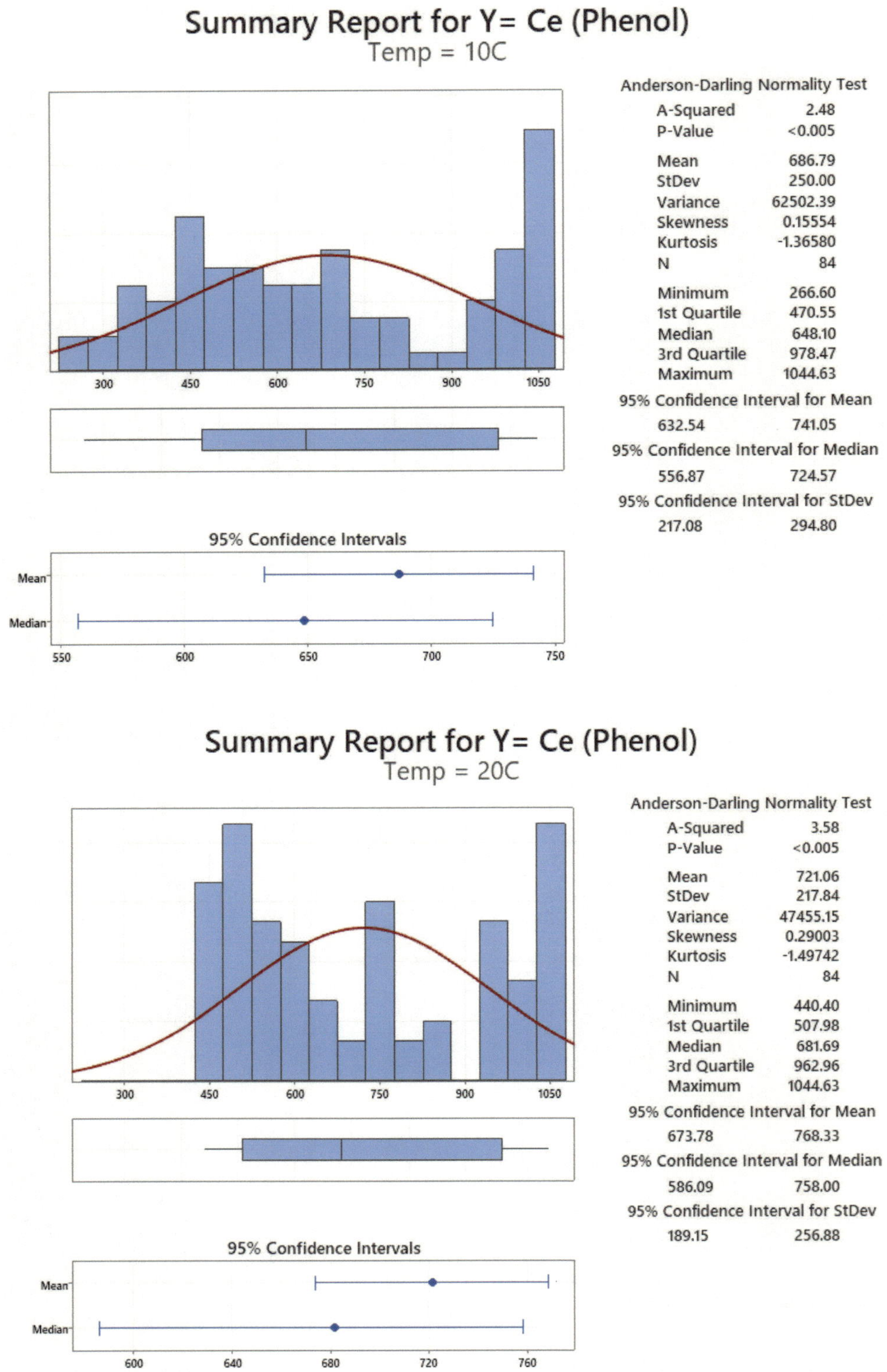

Figure 7.
Summary report for Y=Ce (Phenol) at both temperature values 10 and 10 °C.

The mixed effect of X1 (time) and X2 (mass of peat) on the response was also verified for all the compounds, and two examples of results are shown in **Tables 3** and **4**. The two parameters (time and mass of peat) are significant.

Method					
Variance estimation			Restricted maximum likelihood		
DF for fixed effects			Kenward-Roger		
Rows unused			4		
Factor information					
Factor		**Type**	**Levels**	**Values**	
X1 (time in h)		Random	7	0; 2; 4; 6; 12; 24; 48	
X2 (m of peat in g)		Random	6	0; 2; 5; 10; 15; 20	
Variance components					
Source	**Var**	**% of Total**	**SE Var**	**Z-Value**	**P-Value**
X1 (time in h)	353913E+04	42.21%	213561E+04	1,657,196	0.049
X2 (m of peat in g)	389073E+04	46.40%	254725E+04	1,527,426	0.063
Error	9550,518,041	11.39%	2465,933,155	3,872,983	0.000
Total	838491E+04				
–2 Log likelihood = 531,627,566.					
Model summary					
S	**R-sq**	**R-sq(adj)**	**AICc**	**BIC**	
97,7268	90.40%	90.40%	538,28	542,77	
Coefficients					
Term	**Coef**	**SE Coef**	**DF**	**T-Value**	**P-Value**
Constant	669,146,032	108,479,706	10.13	6,168,398	0.000
Conditional fits and diagnostics for unusual observations					
Obs	**Y = Ce (Phenol)**	**Fit**	**Resid**	**Std Resid**	
1	1044,633,333	1356,324,767	–311,691,433	-3,747,158	R
40	1044,633,333	862,186,020	182,447,314	2,193,383	R

Table 3.
Mixed Effects Model: Y = Ce (Phenol) versus X1 (time in h); X2 (m of peat in g).

Method			
Variance estimation			Restricted maximum likelihood
DF for fixed effects			Kenward-Roger
Rows unused			4
Factor information			
Factor	**Type**	**Levels**	**Values**
X1 (time in h)	Random	7	0; 2; 4; 6; 12; 24; 48
X2 (m of peat in g)	Random	6	0; 2; 5; 10; 15; 20

Variance components

Source	Var	% of Total	SE Var	Z-Value	P-Value
X1 (time in h)	269547E+04	34.15%	164843E+04	1,635,176	0.051
X2 (m of peat in g)	424531E+04	53.78%	277128E+04	1,531,893	0.063
Error	9528,449,672	12.07%	2460,235,126	3,872,983	0.000
Total	789362E+04				

–2 Log likelihood = 530,425,388.

Model summary

S	R-sq	R-sq(adj)	AICc	BIC
97,6138	89.80%	89.80%	537,07	541,57

Coefficients

Term	Coef	SE Coef	DF	T-Value	P-Value
Constant	610,765,873	105,608,027	9.26	5,783,328	0.000

Conditional fits and diagnostics for unusual observations

Obs	Y=Ce (2,4-Dimethylphenol)	Fit	Resid	Std Resid	
1	965,566,667	1267,401,686	–301,835,019	–3,629,195	R
33	965,566,667	772,406,441	193,160,226	2,322,515	R
40	965,566,667	746,623,612	218,943,055	2,632,521	R

Table 4.
Mixed Effects Model: Y=Ce (2,4-Dimethylphenol) versus X1 (time in h); X2 (m of peat in g).

The regression model for all the compounds with the four parameters is described by **Figures 8–14**, and the mixed effects model for the response with the four parameters is described by **Figure 15**. One-way ANOVA and equal variance test for the studied compounds are described by **Figures 16** and **17**.

The typical surface response obtained is shown in **Figure 18**.

The parameter interactions verified by Design Expert software confirm that the parameters X1 (time) and X2 (mass of peat) are significant on the sorption capacity response. P value is low, less than 0.05 for all the response compounds. Results are presented on **Figure 19**.

The results obtained for all the adsorbed phenolic compounds show that when the contact time is reduced (2 h), the increase in peat mass from 2 μg/L to 20 μg/L does not induce a large variation in the adsorption capacity; on the other hand, at time of longer contact (24 h), the increase in mass significantly from 2 μg/L to 20 μg/L promotes the adsorption of all the phenolic compounds (38.9% (2 h) to 59.2% (24 h) for phenol, 56.1% (2 h) to 62% (24 h) for 4-nitrophenol, 55.2% (2 h) to 76.7% (24 h) for

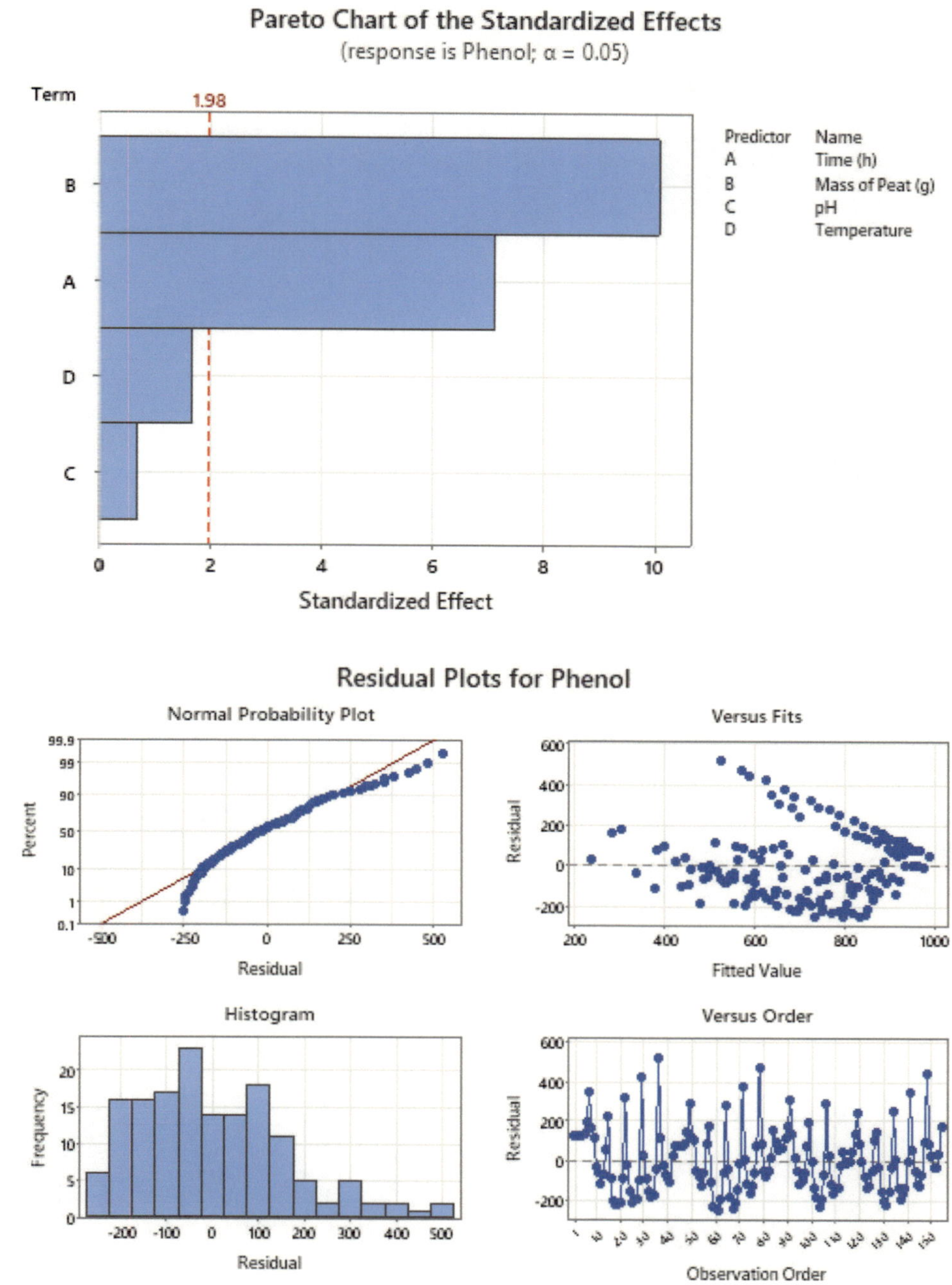

Figure 8.
Regression Analysis: Ce (Phenol) versus Time (h); Mass of Peat (g); pH; Temperature.

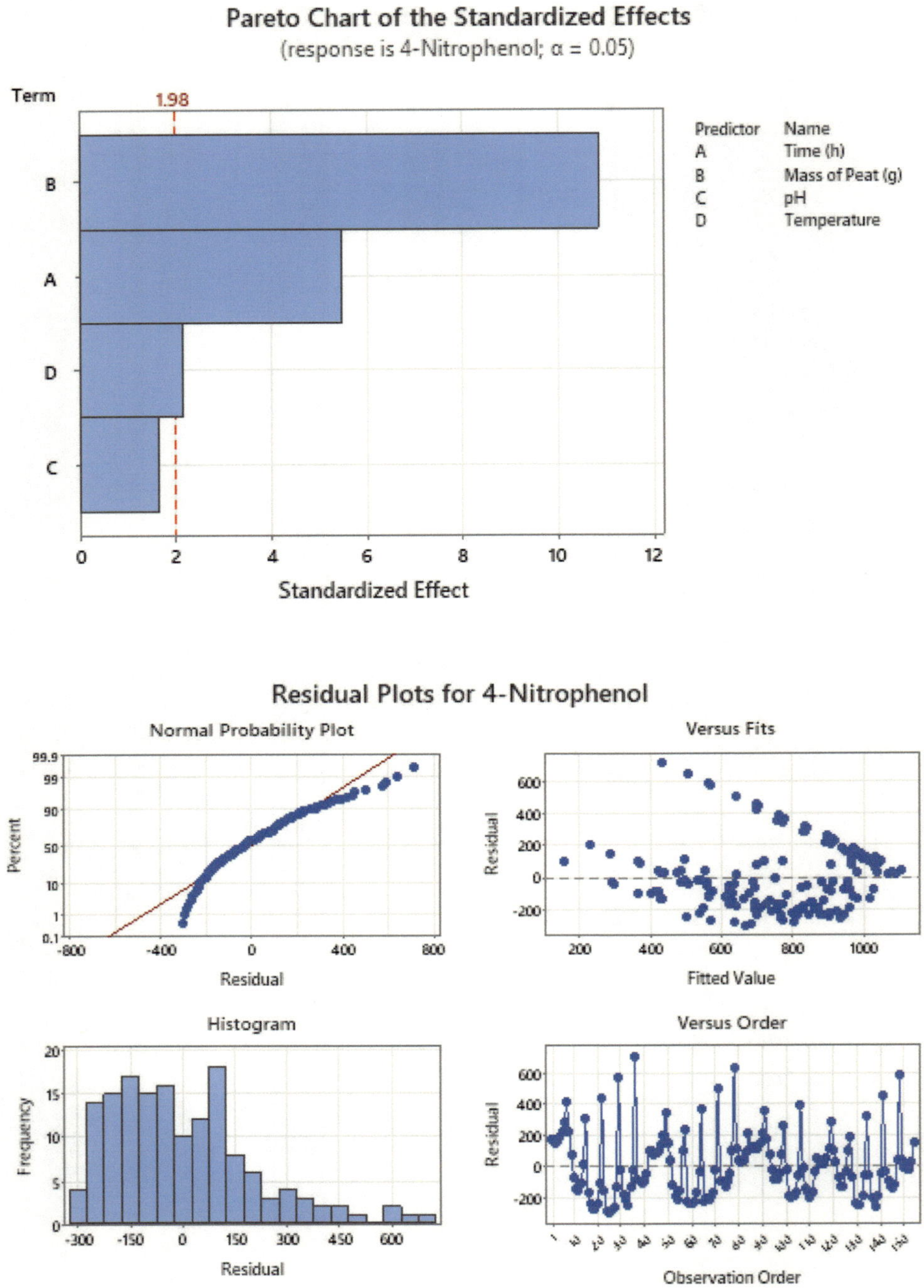

Figure 9.
Regression Analysis: Ce (4-Nitrophenol) versus Time (h); Mass of Peat (g); pH; Temperature.

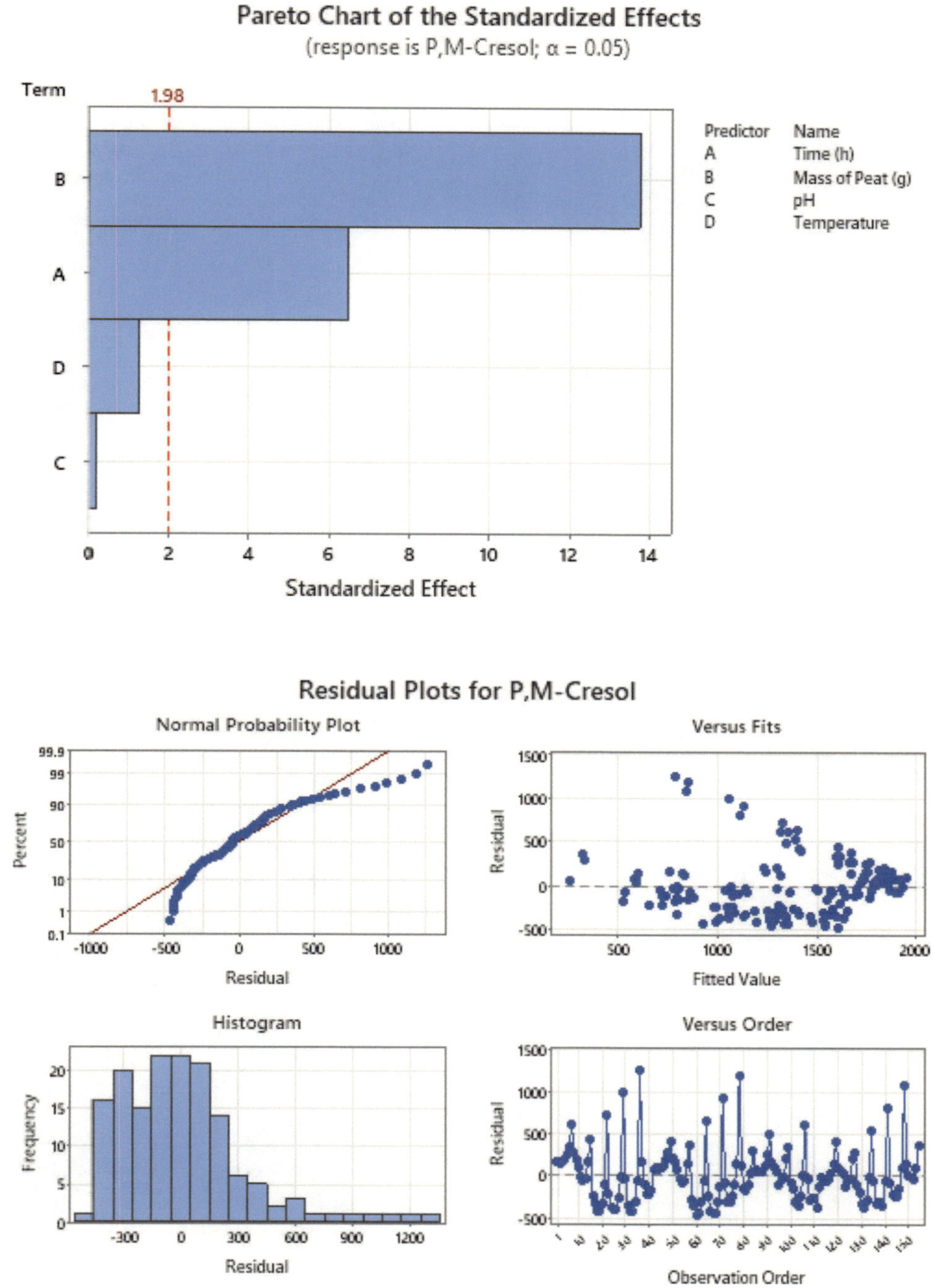

Figure 10.
Regression Analysis: Ce (P,M-Cresol) versus Time (h); Mass of Peat (g); pH; Temperature.

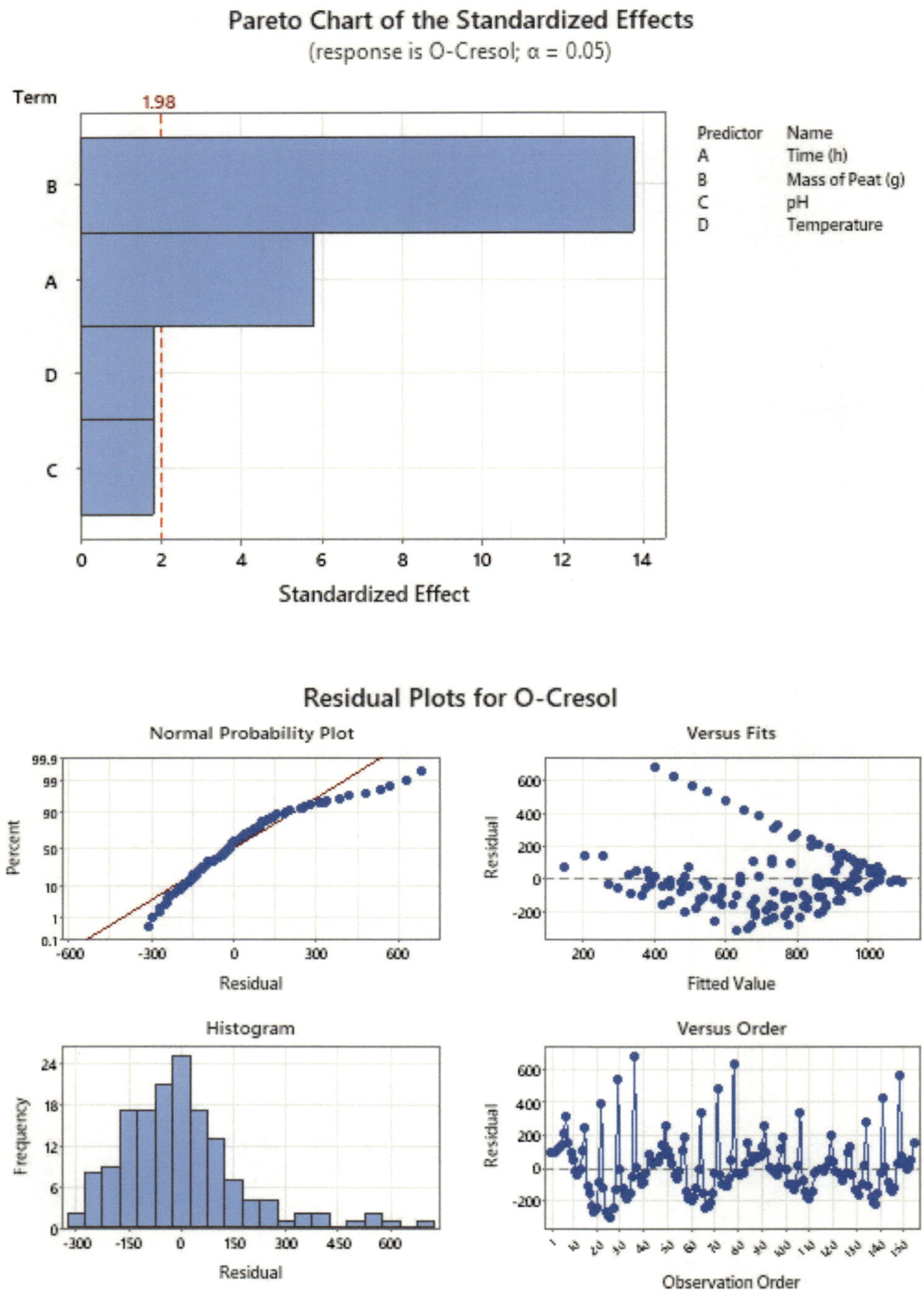

Figure 11.
Regression Analysis: Ce (O-Cresol) versus Time (h); Mass of Peat (g); pH; Temperature.

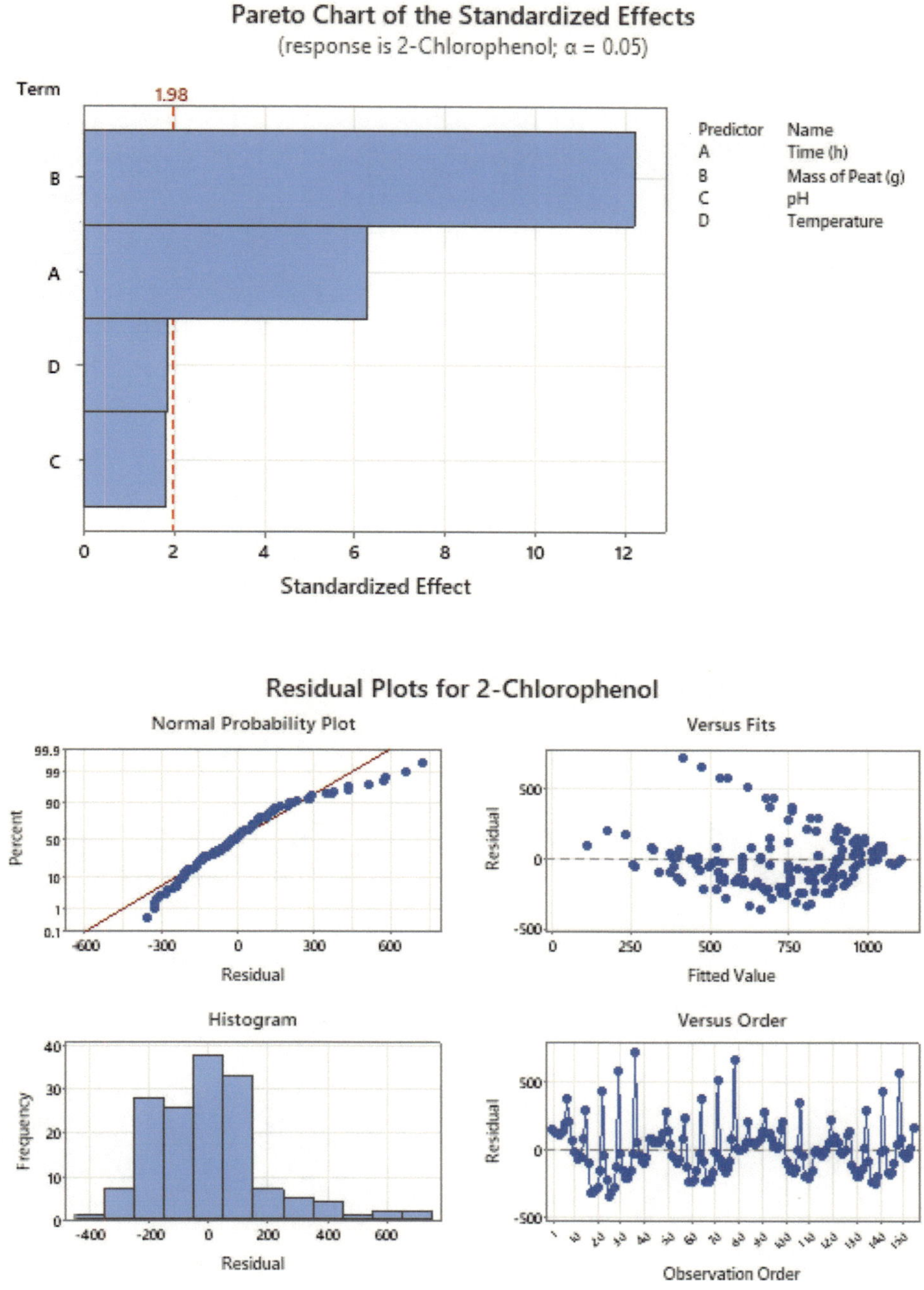

Figure 12.
Regression Analysis: Ce (2-Chlorophenol) versus Time (h); Mass of Peat (g); pH; Temperature.

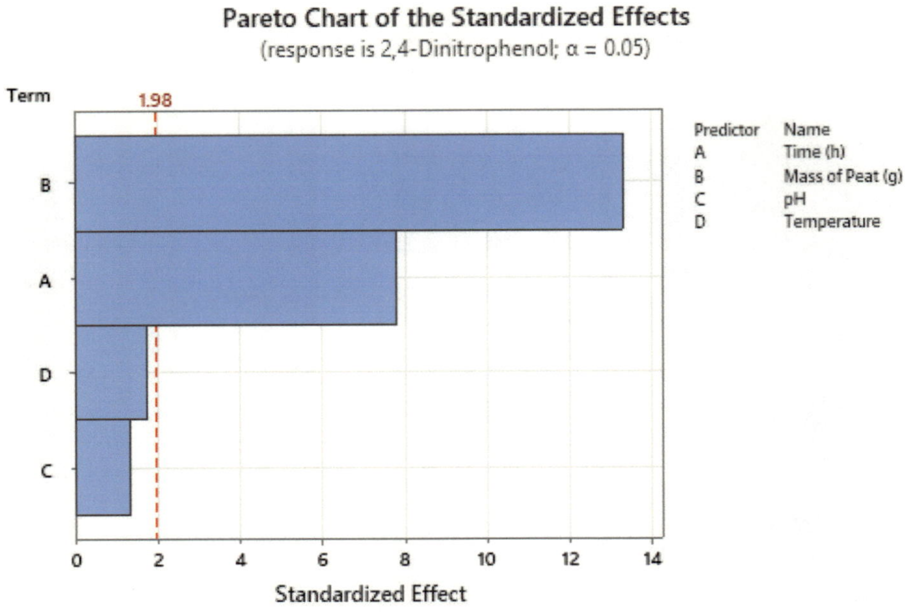

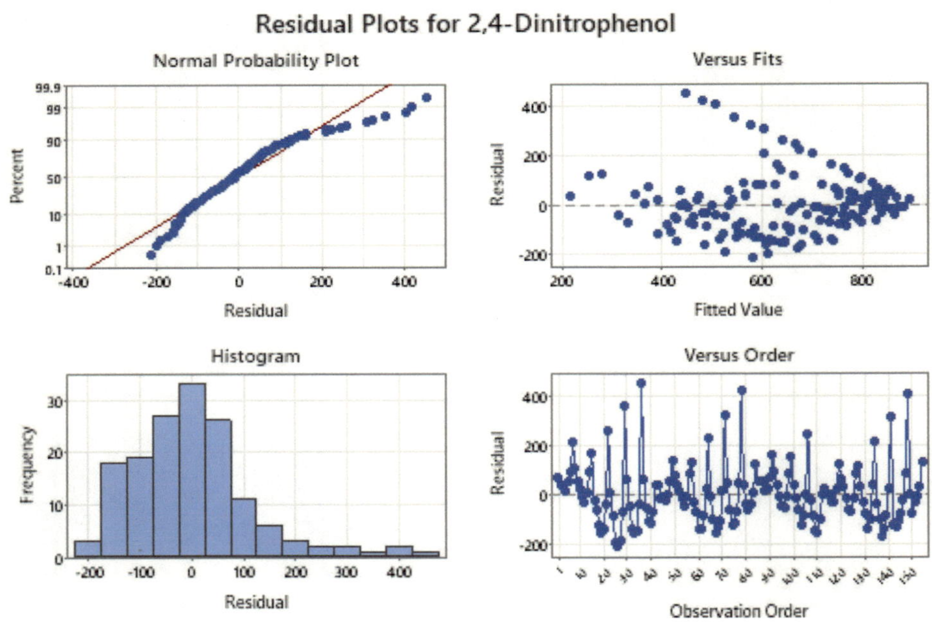

Figure 13.
Regression Analysis: Ce (2,4-Dinitrophenol) versus Time (h); Mass of Peat (g); pH; Temperature.

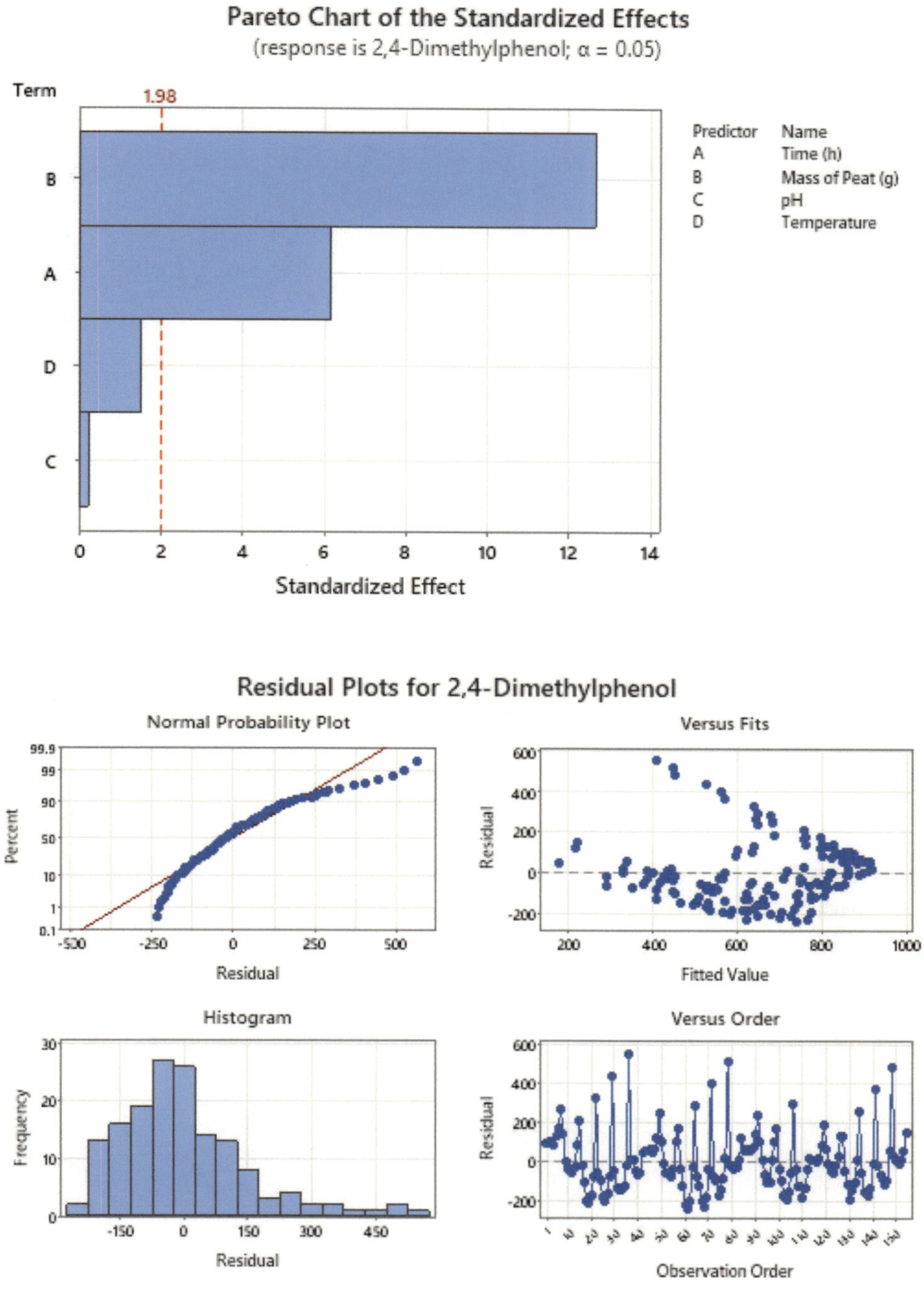

Figure 14.
Regression Analysis: Ce (2,4-Dimethylphenol) versus Time (h); Mass of Peat (g); pH; Temperature.

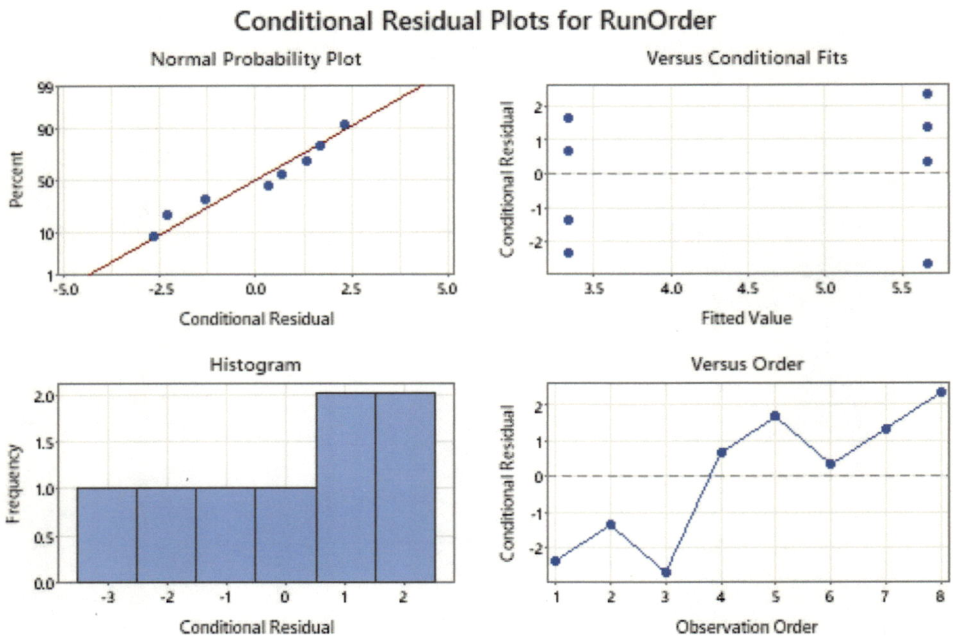

Figure 15.
Mixed Effects Model: RunOrder versus Time (h); Mass of Peat (g); pH; Temperature.

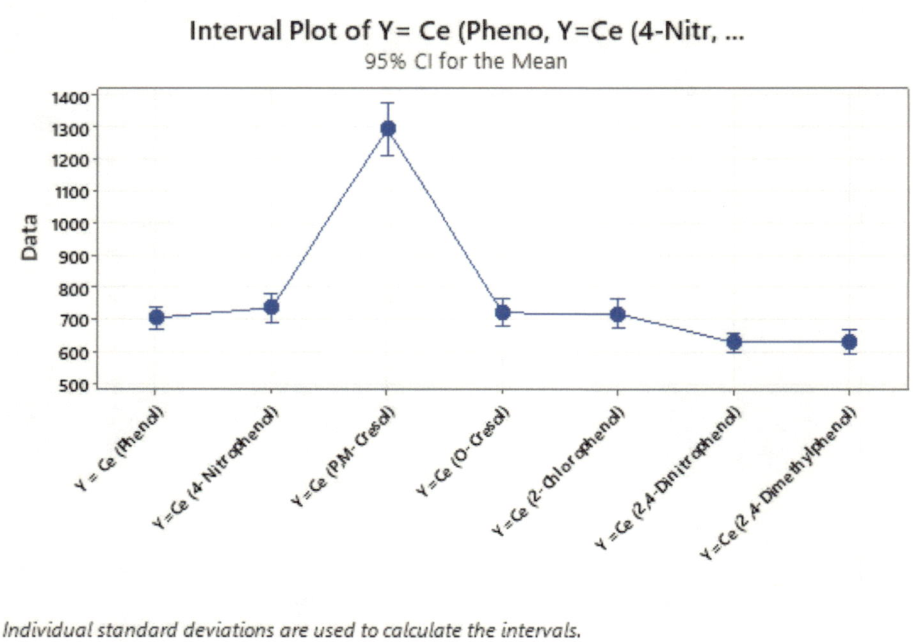

Figure 16.
One-way ANOVA: Y = Ce (Phenol), Y=Ce (4-Nitrophenol), Y=Ce (P,M-Cresol), Y=Ce (O-Cresol), Y=Ce (2-Chlorophenol), Y=Ce (2,4-Dinitrophenol), Y=Ce (2,4-Dimethylphenol).

DOI: http://dx.doi.org/10.5772/intechopen.1008471

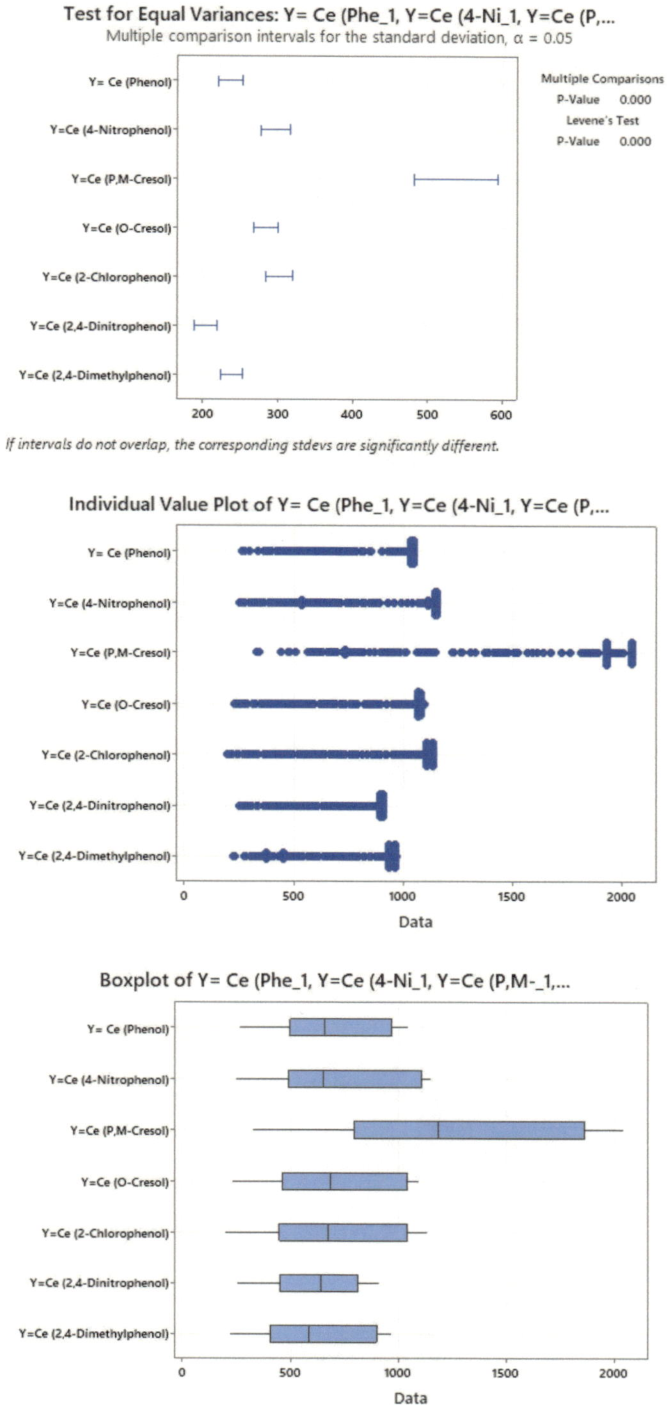

Figure 17.
Test for Equal Variances: Y = Ce (Phenol), Y=Ce (4-Nitrophenol), Y=Ce (P,M-Cresol), Y=Ce (O-Cresol), Y=Ce (2-Chlorophenol), Y=Ce (2,4-Dinitrophenol), Y=Ce (2,4-Dimethylphenol).

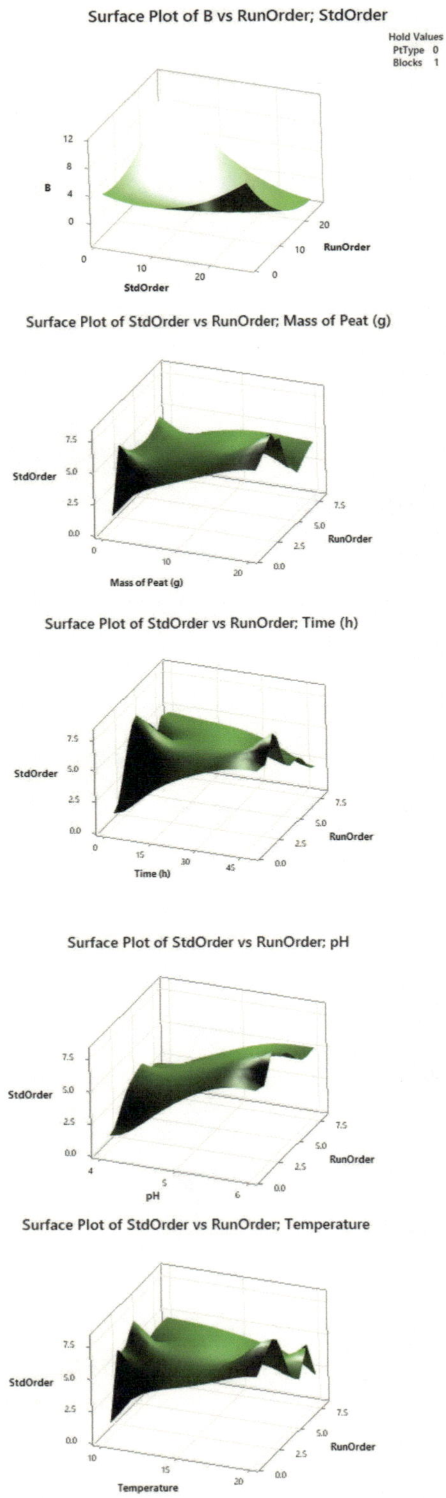

Figure 18.
Typical surface response obtained for all the compounds.

DOI: http://dx.doi.org/10.5772/intechopen.1008471

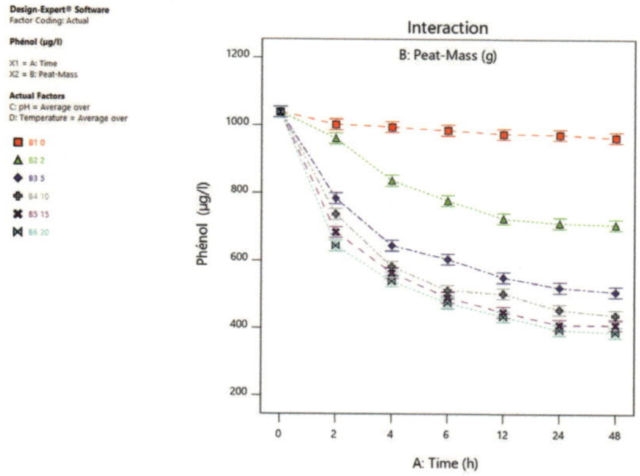

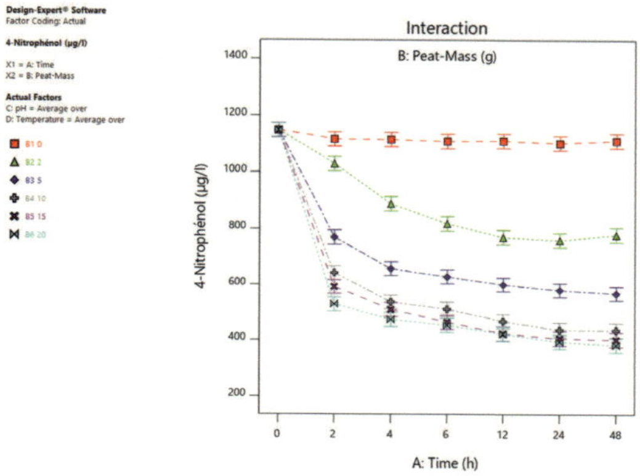

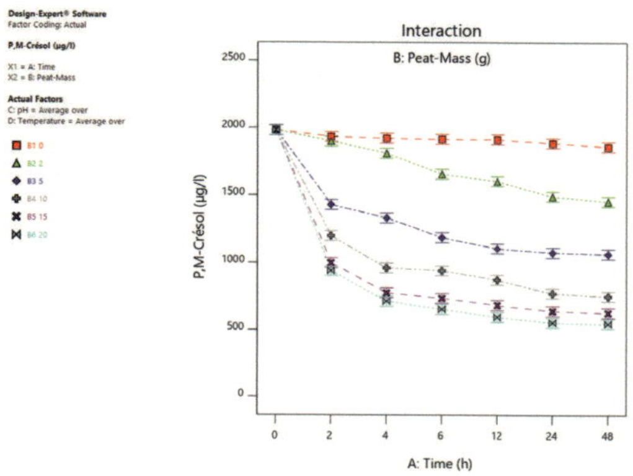

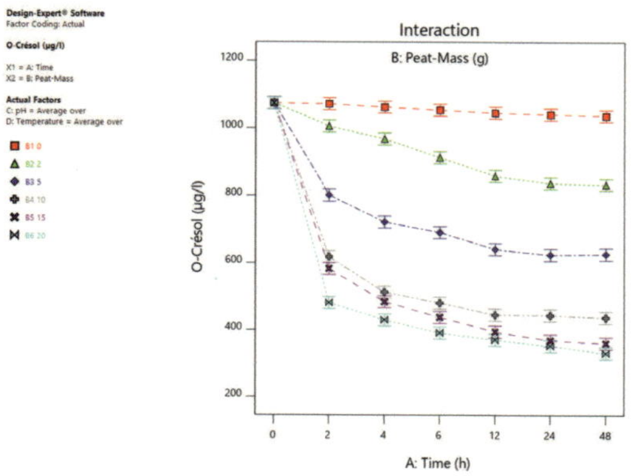

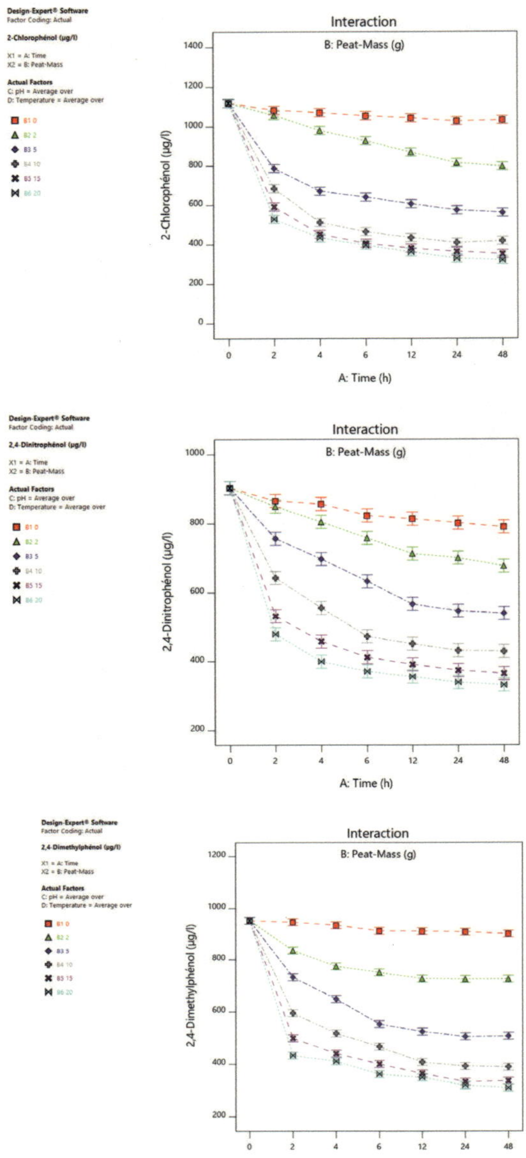

Figure 19.
Parameters interactions study carried out by Design Expert Software.

P.M Cresol, 59.5% (2 h) to 70.4% (24 h) for O-cresol, 56.7% (2 h) to 71.9% (24 h) for 2-chlorophenol and 46.9% (2 h) to 61% (24 h) for 2.4 dinitrophenol).

4. Conclusions

This study contributes to clarify sorption mechanism and kinetics of phenolic compounds in woodwaste leachate by peat media and showed the potential of use of this natural media to remove these compounds. The kinetic study showed that the maximum sorption capacity reached between 20 and 24 h at 10°C and between 16 and

20 h at 20°C. However, it is during the first hours that the sorption process is high. The maximum sorption capacity was evaluated at 68.5 mg/kg (57.87% of the initial concentration) for the most polar compounds: 4-nitrophenol, phenol, and 2-chlorophenol and at 35.2 mg/kg of peat for the least polar compounds such as 2,4-dimethylphenol under conditions of pH 4 and at 10°C. Mixed effect study and interactions parameters verification show that the most significant parameters on the response are time and mass of peat moss (P value less than 0.05 for all the selected compounds). The results of this study confirm that nearly 60% of the most polar compounds initial concentration can be reduced. The great potential of this media can be valuated when combined with another mechanism such as biological process in the same technology.

Acknowledgements

The authors thank the Quebec area sawmill team for the great collaboration, the Natural Sciences and Engineering Research Council of Canada and Investissement Quebec (formaly Centre de Recherche Industrielle du Quebec) for their support. Parts of this chapter were previously published in the doctoral thesis by the same author.

Author details

Najat Kamal[1]*, Rosa Galvez[1], Gerardo Buelna[2] and Abdelaziz Bacaoui[3]

1 Department of Civil Engineering, Laval University, Canada

2 Direction of Environment, Centre de Recherche Industrielle du Québec, Investissement Québec, Canada

3 Department of Chemistry, LCAB, University of Cadi Ayyad, Morocco

*Address all correspondence to: najat.kamal@gmail.com

References

[1] Direction générale des politiques de l'eau. Lignes directrices sur l'industrie du sciage et des matériaux dérivés du bois. Québec: Développement durable, Environnement et Lutte contre les changments climatiques; 2015

[2] Ministère de l'Environnement, du ministère de l'Environnement, de la Faune et des Parcs. (2023). Lignes Directrices sur l'industrie du sciage et des matériaux dérivés du bois. Québec: © Gouvernement du Québec, 2023

[3] Dia O, Drogui P, Buelna G, Dubé R. Strategical approach to prevent ammonia formation during electrocoagulation of landfill leachate obtained from a biofiltration process. Separation and Purification Technology. 2017;**189**:253-259

[4] Dubé R, Turgeon N. Traitement simultané des lixiviats et des gaz d'enfouissement, une approche innovante. Vecteur Environnement. 2014:58-64

[5] Tao W, Hall K, Masbough A, Frankowski K, Duff S. Characterization of leachate from a Woodwaste pile. Water Quality Research Journal. 2005;**40**(4):476-483

[6] Kamal N, Galvez R, Buelna G, Dubé R. Phenolic compounds removal in Woodwaste leachate by a trickling biofilter. In: Em Phenolic Compounds-Natural Sources, Importance and Applications. London, UK: INTECH Publishing; 2017. pp. 373-393 ISBN 978-953-51-2958-5

[7] Bahmani R, Kim D, Modareszadeh M, Thompson AJ, Park JH, Yoo HH, et al. The mechanism of root growth inhibition by the endocrine disruptor bisphenol a (BPA). Environmental Pollution (Barking, Essex: 1987. 2020;**257**:113516

[8] de Oliveira KM, de Sousa Carvalho EH, Filho RD, Wilke Sivek T, Thá EL, de Souza IR, et al. Single and mixture toxicity evaluation of three phenolic compounds to the terrestrial ecosystem. Journal of Environmental Management. 2021;**296**:113226

[9] Hadjar H, Hamdi B, Bachiller-Baeza B, Doña-Rodríguez J. Efficient sorption performance of carbon-diatomaceous silica compounds towards phenol. Surfaces and Interfaces. 2021;**24**:101101. DOI: 10.1016/j.surfin.2021.101101. ISSN 2468-0230

[10] Kim D, Kwak J, An Y. Physiological response of crop plants to the endocrine-disrupting chemical nonylphenol in the soil environment. Environmental Pollution. 2019;**251**:573-580

[11] Wang P, Tyndall S, Rahman T, Roy P, Jahromi H, Adhikari S, et al. Sorption and recovery of phenolic compounds from aqueous phase of sewage sludge hydrothermal liquefaction using bio-char. Chemosphere: Part 1. 2022;**287**:131934

[12] Mohammadi S, Kargari A, Sanaeepur H, Abbassian K, Najafi A, Mofarrah E. Phenol removal from industrial wastewaters: A short review. New pub Balaban. 2015;**53**:2215-2234

[13] Wang C, Xu Y, Shi Y, Wang C, Duan L, Gu W, et al. Verification on the developmental toxicity of short-term exposure to phenol in rats. Biomedical and Environmental Sciences. 2020;**33**(6):403-413

[14] Browne P, Van Der Wal L, Gourmelon A. OECD approaches

and considerations for regulatory evaluation of endocrine disruptors. Molecular and Cellular Endocrinology. 2020;**504**:110675

[15] Pham N, Huynh T, Nasir M. Environmental consequences of population, affluence and technological progress for European countries: A Malthusian view. Journal of Environmental Management. 2020;**260**:110143

[16] Lopez-Munoz M, Sotto A, Arsuaga J, Van der Bruggen B. Influence of membrane, solute and solution properties on the retention of phenolic compounds in aqueous solution by nanofiltration membranes. Separation and Purification Technology. 2009;**66**:194-201

[17] Vieira V, Prieto MA, Barros L, Coutinho JA, Ferreira IC, Ferreira O. Enhanced extraction of phenolic compounds using choline chloride based deep eutectic solvents from juglans regia l. Industrial Crops and Products. 2018;**115**:261-271

[18] Zilnik L, Jazbinsek A. Recovery of renewable phenolic fraction from pyrolysis oil. Separation and Purification Technology. 2012;**86**:157-170

[19] Antonopoulou M, Papadopoulos V, Konstantinou I. Photocatalytic oxidation of treated municipal wastewaters for the removal of phenolic compounds: Optimization and modeling using response surface methodology (RSM) and artificial neural networks (ANNs). Journal of Chemical Technology and Biotechnology. 2012;**87**(10):1385-1395

[20] Mohite RG, Garg A. Performance of heterogeneous catalytic wet oxidation for the removal of phenolic compounds: Catalyst characterization and effect of pH, temperature, metal leaching and non-oxidative hydrothermal reaction. Journal of Environmental Chemical Engineering. 2017;**5**(1):468-478

[21] Karami T, Elyasi S, Amani T. Modeling and optimizing of electrocoagulation process in treating phenolic wastewater by response surface methodology: Precise evaluation of significant variables. International journal of Environmental Science and Technology. 2018;**15**:2389-2398

[22] Chezeau B, Vial C. Chapter 19- Modeling and simulation of the biohydrogen production processes. In: Pandey A, Mohan SV, Chang J-S, Hallenbeck PC, Larroche C, editors. Biomass, Biofuels, Biochemicals. 2nd ed. Biohydrogen; 2019. pp. 445-483

[23] Klonowska A, Gaudin C, Fournel A, Asso M, Le Petit J, Giorgi M, et al. Characterization of low redox potential laccase from the basidiomycete C30. European Journal of Biochemistry. 2002;**2002**(269):6119-6125

[24] Kumaran P, Paruchuri YL. Kinetics of phenol biotransformation. Water Research. 1997;**31**(1):11-22

[25] Mohanty SS, Jena HM. Biodegrdation of phenol by free and immobilized cells of a novel pseudomonas sp. NBM11. Brazilian Journal of Chemical Engineering. 2017;**34**(1):75-84

[26] Tabis B, Skoneczny S. Diffusional penetration depths in biofilms. Chemical Engineering and Processing. 2010;**31**(4):857-871

[27] Halvorson JJ, Gonzalez JM, Hagerman AE, Smith JL. Sorption of tannin and related phenolic compounds and effects on soluble-N in soil. Soil Biology and Biochemistry. 2009;**41**(9):2002-2010

[28] Hsieh CT, Teng H. Influence of mesopore volume and adsorbate size on adsorption capacities of activated carbons in aqueous solutions. Carbon. 2000;**38**(6):863-869

[29] Mishra P, Singh K, Dixit U. Adsorption, kinetics and thermodynamics of phenol removal by ultrasound-assisted sulfuric acid-treated pea (Pisum sativum) shells. Sustainable Chemistry and Pharmacy. 2021;**22**:100491

[30] Ou Y, Chang Y, Lin F, Chang M, Yang C, Shih Y. Competitive sorption of bisphenol a and phenol in soils and the contribution of black carbon. Ecological Engineering. 2016;**92**:270-276

[31] Peñas FJ, Romo A, Isasi JR, San José MJ, Alvarez S. Kinetic modeling of sorption–desorption cycles for phenol removal with a cyclodextrin polymer. Journal of Industrial and Engineering Chemistry. 2019;**75**:93-99

[32] Sotto A, Arsuaga JM, Van der Bruggen B. Sorption of phenolic compounds on NF/RO membrane surfaces: Influence on membrane performance. Desalination. 2013;**309**:64-73

[33] Viraraghavan T, Alfaro F. Adsorption of phenol from wastewater by peat, fly ash and bentonite. Journal of Hazardous Materials. 1998;**57**:59-70

[34] Wu W, Lan Y, Zeng Y, Lin D, Yang K. Nonlinear sorption of phenols and anilines by organobentonites: Nonlinear partition and space limitation for partitioning. Science of the Total Environment. 2020;**736**:139609

[35] Xing B, McGill WB, Dudas MJ, Maham Y, Helper L. Sorption of phenol by selected biopolymers: Isotherms, energetics, and polarity. Environmental Science and Technology. 1994;**28**(3):466-473

[36] Kamal N, Galvez R, Buelna G. Application of a solid phase extraction liquid chromatography to quantify phenolic compounds in woodwaste leachate. Water Quality Research Journal of Canada. 2014a;**49**:210-222

[37] Kamal N, Galvez R, Buelna G. Application of a solid phase extraction-liquid chromatography method to quantify phenolic compounds in woodwaste leachate. Water Quality Research Journal of Canada. 2014b;**49**:210-222

Chapter 3

A Fixed-Bed Column Sorption: Breakthrough Curves Modeling

Andrés A. Abin-Bazaine, Mario A. Olmos-Marquez and Alfredo Campos-Trujillo

Abstract

Global attention has increasingly focused on environmental pollution due to its widespread and devastating impact. The urgency of addressing climate change has propelled it to the forefront of governmental agendas worldwide, emphasizing the need for actions to secure a pollution-free future. Pollution treatment methods have consequently gained global significance, with adsorption emerging as a particularly relevant approach, especially in developing economies. Adsorption proves to be a cost-effective, safe, efficient, and easily manageable method that can utilize low-cost or waste materials. In designing treatment systems based on adsorption, batch tests are crucial, employing adsorption isotherms such as Langmuir and Freundlich to understand the phenomenon. While equilibrium points are essential in some situations, continuous processes benefit from column implementations, where a fundamental understanding of breakthrough curves becomes pivotal. Various adsorption kinetic models, such as the Thomas model, Adams–Bohart model, Yoon–Nelson model, and bed-depth/service time (BDST) model, explain and determine breakthrough curves. The assessment of these models for compatibility with experimental data and model-generated data is essential. Criteria such as Mean Relative Error (MRE) and Normalized Relative Mean Square Error (NRMSE) are commonly employed to objectively select the most suitable model for a given scenario.

Keywords: fixed-bed column adsorption, isotherms sorption, adsorbent, adsorbate, process parameters

1. Introduction

Bois-Reymond initially proposed the concept of "adsorption", but was later introduced to the scientific community by Kayser. It is defined as the accumulation of a specific compound at the interface between two phases, this phenomenon involves the transfer from one phase to the other and subsequently adheres to the surface. This process is considered intricate and largely influenced by the surface chemistry or the characteristics of the sorbent, sorbate, and the environmental conditions between the two phases [1]. This procedure can be efficiently employed to move contaminants or pollutants from wastewater and concentrate them on the surface of adsorbents [2]. Certain compounds undergo transportation from one phase to another before adhering to a surface. This process is regarded as a complex phenomenon primarily reliant on the surface chemistry or characteristics of the sorbent and sorbate, as well as the system conditions between the two phases [1].

The material that gets adsorbed onto the surface of another substance is termed the adsorbate. Meanwhile, the substance existing in bulk, where this adsorption occurs on the surface, is known as the adsorbent. This interaction can manifest at interfaces such as liquid-liquid, liquid-solid, gas-liquid, or gas-solid. Among these types, it is primarily liquid-solid adsorption that finds extensive application in water and wastewater treatment [1]. The process consists of a sequence of four steps, during which the dissolved substance (adsorbate) undergoes transitions to reach the boundary layer and adheres to the adsorbent. The following four stages describe the movement of the solute (adsorbate) toward the boundary layer and its binding to the adsorbent:

i. Advective transport: Dissolved particles are transported from the overall solution to an immobilized film layer through advective flow, axial dispersion, or diffusion.

ii. Film transfer: Solute particles permeate the immobile water film layer and accumulate within it.

iii. Mass transfer: Solute particles gather on the surface of the adsorbent.

iv. Intraparticle diffusion: The solute undergoes movement into the pores of the adsorbent [1].

Sorption is initiated by the concentration difference between the liquid and the surface of the adsorbing solid. During this phase, the movement of solute molecules is solely guided by molecular diffusion across the interface. The functional groups within the adsorbent solid determine the affinity and the type of interaction with the adsorbed molecules. The binding of these molecules to the adsorbent causes a change in entropy, specifically, a decrease in entropy [2].

There exist two primary types of adsorption: Physical adsorption is instigated by weak Van der Waals forces of attraction. This type of adsorption is reversible and is distinguished by low enthalpy values, typically around 20 kJ mol^{-1}. In physical adsorption, the attractive forces between the adsorbed molecules and the solid surface are feeble, allowing the molecules to move freely across the surface rather than being firmly attached to a specific location on the adsorbent surface. Electrostatic forces, such as dipole-dipole interactions, dispersion forces, and hydrogen bonding, play a role in the interactions between the adsorbate and adsorbent in physical adsorption [1, 2].

On the other hand, chemisorption involves chemical bonding between the sorbate and sorbent molecules. This type of sorption is irreversible and exhibits a higher enthalpy of sorption, around 200 kJ mol^{-1}. In chemisorption, the attraction between the sorbent and sorbate is pivotal, and this is facilitated by more robust electrostatic forces, including covalent or electrostatic chemical bond [1, 2]. The ranges of energy for each reactions are: (II) Van der Waals force 4 < enthalpy <10 kJ mol^{-1}, (II) Hydrophobic Force <5 kJ mol^{-1}, (II) Dipole Force 2 < enthalpy <29 kJ mol^{-1}, (IV) Hydrogen Bond 2 < enthalpy <40 kJ mol^{-1}, (V) Coordination Exchange 40 < enthalpy <60 kJ mol^{-1}, and (VI) Chemical Bond Enthalpy >60 kJ mol^{-1} [1, 2]. The accumulation of pollutants in the environment is a product of human activities, such as mining and the discharge of industrial waste [3]. One of the most promising technologies for treating large volumes of diluted pollutants in solutions is sorption [4]. Sorption describes a mass transfer process, where a dissolved material is transferred directly to the surface of the solid phase, after it is bound by chemical and/or physical interactions [5]. The sorption capacity of a material obtained from batch studies presents useful results to know the effectiveness of a sorption system, however, this information is generally not applicable to most treatment systems where the time of contact is not enough to reach equilibrium, therefore, there is a need to perform equilibrium studies using columns [3]. The batch operation mode is straightforward, employing a consistent volume approach, making it simple to operate. Conversely, the continuous flow operation, with varying volumes, is somewhat more complex, necessitating additional devices like pumps, flow regulators, strainers, temperature controllers, and so on. Batch processing is suitable for initial research endeavors to explore isotherms and kinetics. Meanwhile, continuous flow studies delve into a comprehensive understanding of the dynamics of biosorption processes during large-scale operations [2]. The fixed-bed adsorption column systems allow to determine the maximum performance of the adsorbent materials and to identify the best dynamic operating conditions (6). Fixed-bed column adsorption has many advantages due to its simple operation and high removal efficiency [6]. Understanding breakthrough curves is essential in designing a continuous flow system for the adsorption of pollutants. Typically, this information is obtained through experimental determination or modeling, relying on kinetic behavior and the isothermal model [7].

The continuous operation of a fixed-bed column adsorption process serves as a crucial link between laboratory experimentation and real-world applications, offering valuable insights. Nevertheless, optimizing operational parameters through direct column experimentation, specifically with limited data points, can be an arduous, time-consuming, and costly endeavor. Hence, numerous mathematical models have been devised to predict actual adsorption behaviors and deliver effective design insights for columns, negating the need for physical experimentation [8]. Adsorption can be conducted in both batch and continuous modes. Batch adsorption experiments are crucial for collecting fundamental data regarding pollutant removal through adsorption. However, to effectively apply this process in wastewater treatment, continuous adsorption studies are indispensable for practical implementation [9]. In general, continuous adsorption is carried out with a fixed-bed column where wastewater encounters a stationary bed of adsorbent. In this continuous adsorption process, the wastewater constantly enters and exits the column, maintaining a dynamic equilibrium. This equilibrium is upheld as the concentration of ions in both the solution and the adsorbent within the column is constantly changing [9].

In experiments involving a fixed-bed column, a specific amount of untreated or pre-treated biosorbents within a solid matrix is densely packed into a column of defined length. The column is equipped with sieves at the bottom to prevent the biosorbent from being carried away by the liquid phase. This loaded column is then placed in a thermostat to maintain a constant temperature and is connected to a positive displacement pump, often a peristaltic pump, with a flow regulator. The liquid phase, containing dissolved substances (contaminants), flows through the fixed-bed column at a predetermined rate.

As the fixed-bed becomes wet, dissolved substances move out of the liquid phase and adhere to the surface of the stationary matrix due to their affinity. This translocation and binding of solute molecules persist until the pores or functional groups of the biosorbents reach equilibrium. Once saturation is attained, further adsorption becomes impractical, and the transfer of solute at the solid-liquid interface ceases. Additionally, the transfer of solutes across the phase boundary encounters diffusion resistance from the surrounding film and within the particles.

The adsorbed molecules necessitate a sufficient amount of time to remain in the column. The residence or retention time is contingent on the volumetric flow rate of the liquid samples and is, therefore, a crucial parameter for the optimal design of the continuous biosorption process [2].

The effectiveness of continuous adsorption is assessed by evaluating parameters such as column efficiency in removing contaminants, column uptake capacity, and breakthrough curve profile. These parameters are influenced by operating factors such as the bed diameter, bed depth, flow rate, and initial concentration of the contaminant [9].

2. Adsorption models for column study

2.1 Calculation of the dynamic adsorption parameters

The performance of an adsorption column can be determined according to breakthrough curves (**Figures 1** and **2**). The duration of breakthrough occurrence and the configuration of the breakthrough curve are crucial attributes for assessing the operation and dynamic response of a sorption column [10, 11].

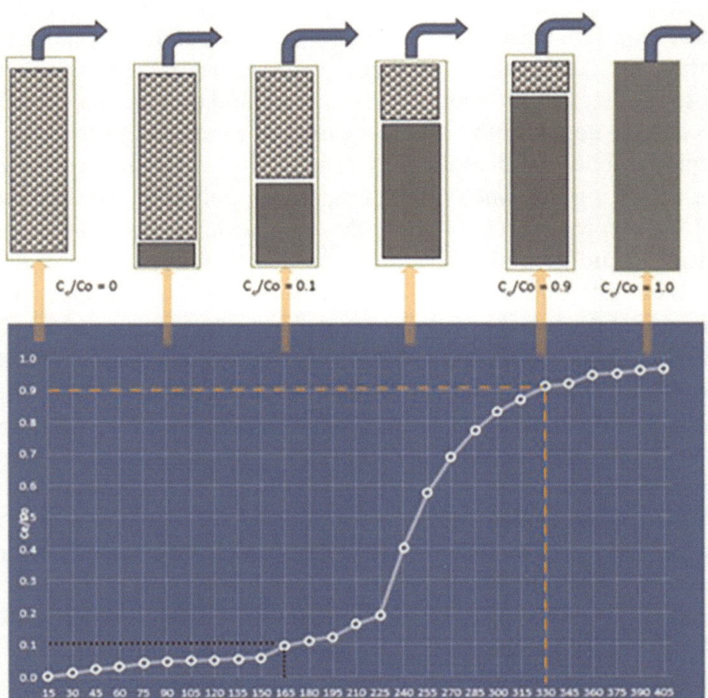

Figure 1.
Representation of breakthrough curve.

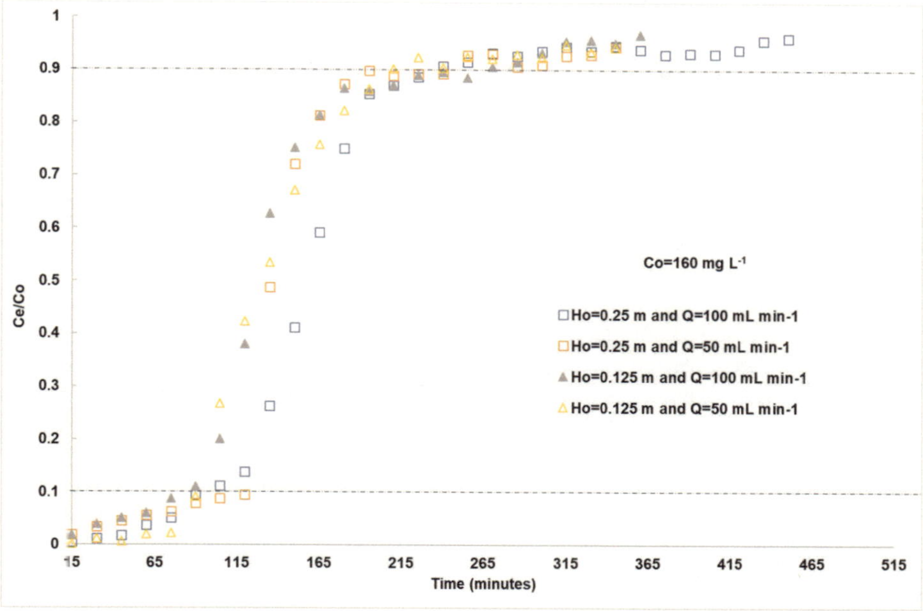

Figure 2.
Behavior of a breakthrough curve, when the initial concentration remains constant, varying the height of the column (H_o) and the flow rate (Q).

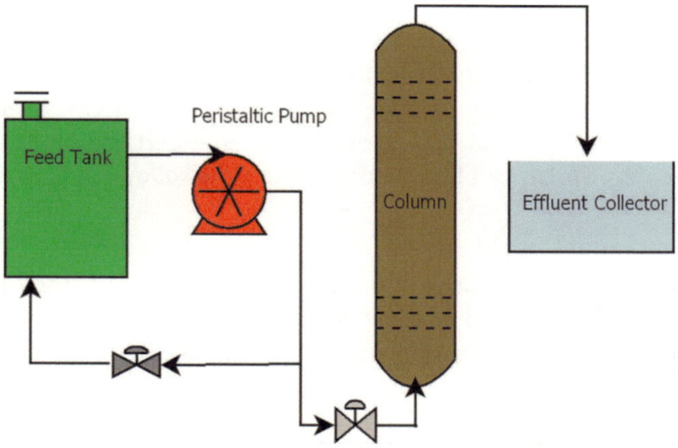

Figure 3.
Schematic illustrator continuous adsorption process.

Figure 3 shows a scheme for a continuous adsorption process. This scheme can be used for a laboratory experiment or for a larger scale such as a pilot plant. To regulate the fed flow, an arrangement of valves can be used to return part of the flow to the feed tank.

Figure 1 shows the diagram of the development of the adsorption process and the representation of the breakthrough curve as well as a diagram of the saturation process of the material with which the column was packed. The darkest part of the column illustrates how the material becomes saturated over time (**Figure 4**).

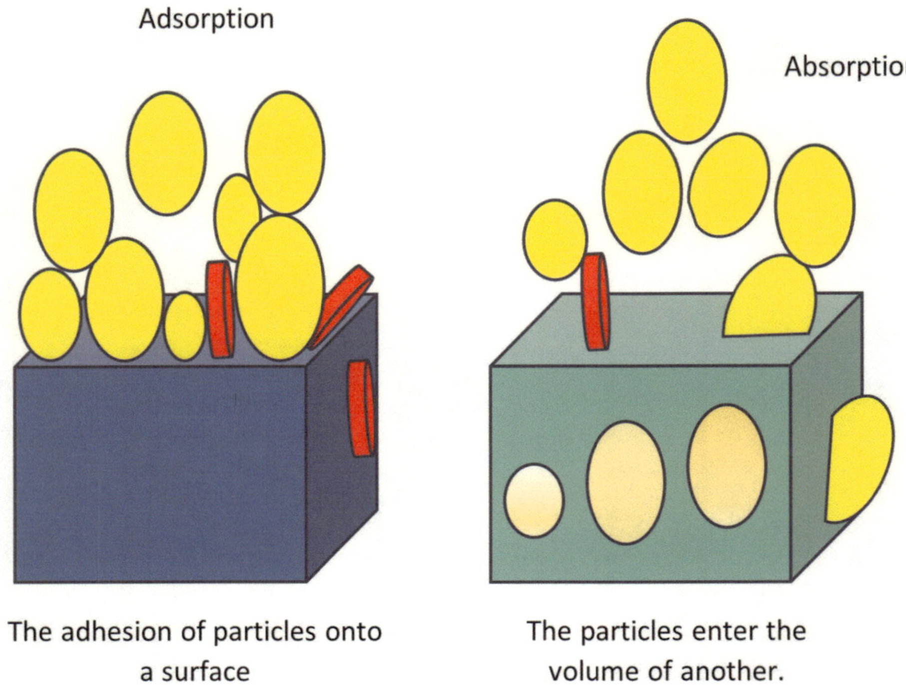

Figure 4.
Scheme of the adsorption process.

The breakthrough point is typically reached when the effluent concentration (C_e) from the column reaches around 5 to 10% of the influent concentration (C_0) [10, 11]. The point where the effluent concentration reaches 90% is usually called the "point of column exhaustion" [10, 11]. By plotting C_e/C_o versus the reaction time, the breakthrough curves for an adsorbent will be obtained under certain conditions. The effluent volume, V_{eff} (mL), can be calculated from the following equation [8, 10, 11]:

$$V_{eff} = Qt_{total} \tag{1}$$

Where: Q is the volumetric flow rate (mL min^{-1}), and t_{total} is the total flow time (min). For a given flow rate and influent concentration, the maximum adsorption capacity of column, q_{total} (mg), is obtained using [8, 10–13]:

$$q_{total} = \frac{QA}{1000} = \frac{Q}{1000}\int_{t=0}^{t=t_{total}} C_t dt \tag{2}$$

$$q_e = \frac{q_{total}}{m} \tag{3}$$

Where: A is the area column cross-section area (cm^2), C_t ion concentration after time t (mg L^{-1}), q_e is maximum capacity of the column (mg of pollutant g^{-1} Adsorbent), m is the mass of adsorbent in the column (g), Q is the volumetric flow rate (mL min^{-1}), and t_{total} is the total flow time (min).

Arranging the Eqs. (2) and (3), obtained [3, 8, 10–14]:

$$q_e = \frac{C_0 Q}{m * 1000}\int_{t=0}^{t=t_{total}}\left(1 - \frac{C_e}{C_0}\right)dt \tag{4}$$

The total quantity of ions entering the column (m_{total}) is computed using the following Eq. (17):

$$m_{total} = \frac{C_0 Q t_{total}}{1000} \tag{5}$$

The total pollutant ions getting into the fixed-bed W (mg) can be calculated from the following equation [8, 12, 14]:

$$W = C_0 F t_e \tag{6}$$

Where: F is the volumetric flow rate (mL h^{-1}) and t_e is the exhaustion time (h).
Total removal percent of ions can be calculated from [12–14]:

$$Y\% = \frac{q_{total}}{W_{total}} * 100 \tag{7}$$

The flow rate represents the empty bed contact time (EBCT) in the column, as described in [10]:

$$EBCT(\min) = \frac{\text{Bed volume (mL)}}{\text{Flow rate (mL/ min)}} \tag{8}$$

Typically, the breakthrough time on the curve is defined as the moment when the contaminant concentration in the effluent reaches the specified limit standard or a predetermined fraction of the initial concentration [8]. The breakthrough time (t_b) refers to the point at which 10% of the initial concentration is detected in the effluent concentration, known as the concentration breakthrough (C_b), and is closely related to the Fraction Bed Utilization (FBU) [6, 15].

$$C_b = 0.1C_0 \tag{9}$$

$$C_S = 0.9C_0 \tag{10}$$

$$FBU = \frac{q_b}{q_s} \tag{11}$$

$$q_b = \frac{C_0\ Q}{1000\ m} \int_0^{t_b} \left(1 - \frac{C_b}{C_0}\right) \tag{12}$$

$$q_s = \frac{C_0 Q}{1000\ m} \int_0^{t_s} \left(1 - \frac{C_s}{C_0}\right) \tag{13}$$

Where: C_b is breakthrough concentration (mg L^{-1}), C_S is saturation concentration (mg L^{-1}), q_b is the amount of ions adsorbed at breakthrough time (mg g^{-1}), q_s is the amount of ions adsorbed at saturation time (mg g^{-1}), t_b is breakthrough time (min), and t_s is saturation time (min).

2.2 Adsorption kinetic models

Effectively designing a column adsorption process necessitates accurately predicting the breakthrough curve in the effluent. Throughout the years, numerous straightforward mathematical models have been created to describe and analyze laboratory-scale column studies, catering specifically to industrial applications [10, 11].

2.2.1 Thomas model

The model assumes that there is no axial diffusion when passing the solution through the filling material of the column and the kinetic model of Pseudo-second order is the one that best describes the adsorption process [11, 13, 16]. The equation that describes it is the following:

$$\frac{C_e}{C_o} = \frac{1}{1 + exp\left[\frac{K_{Th} q_e m}{Q} - K_{Th} C_0 t\right]} \tag{14}$$

Its linear form is expressed as:

$$Ln\left[\frac{C_0}{C_e} - 1\right] = \frac{K_{Th} q_e m}{Q} - K_{Th} C_0 t \tag{15}$$

Where C_e is the effluent ions concentration (mg L^{-1}); C_0 is initial ions concentration (mg L^{-1}); Q is flow rate (L h^{-1}); qe is the equilibrium column capacity (mg g^{-1}); m is the weight of Zeolite in column (g); K_{Th} is the Thomas model constant (mL min^{-1} mg^{-1}), and t stands for total flow time (min).

The values of K_{Th} and q_e can be determined from the linear plot of $Ln[(C_o/C_e)\text{-}1]$ against t [10, 13, 16].

2.2.2 Bohart and Adams model

In this model, it is assumed that the rate of adsorption is directly proportional to both the concentration of the adsorbate and the remaining capacity of the adsorbent for the adsorbate [11]. This model adds that the equilibrium is not instantaneous and, therefore, is used to model the initial part of the breakthrough curves. Based on the surface reaction theory, established a fundamental equation, which describes the relationship between Ce/C_0 and t in a continuous system.

The equation that describes it is the following [11]:

$$\frac{C_e}{C_0} = \frac{\exp(K_{BA} \cdot C_0 \cdot t)}{\exp\left(K_{BA} \cdot q_{max} \cdot \frac{Z}{U_0}\right) - 1 + exp\left(K_{BA} \cdot C_0 \cdot t\right)} \tag{16}$$

The linear expression is as follows [16, 17]:

$$Ln\frac{C_e}{C_0} = K_{AB} C_0 t - K_{AB} N_0 \left(\frac{Z}{U_0}\right) \tag{17}$$

Where: C_0 and Ce are the influent and effluent concentration (mg L^{-1}), respectively; K_{AB} is the kinetic constant (L mg^{-1} min^{-1}), t breakthrough time min, q_{max} maximum adsorption capacity (mg ions g^{-1} absorbent), N_0 is the saturation concentration (mg L^{-1}), Z is the bed depth of the fix-bed column (cm), and U_0 is the superficial velocity (cm min^{-1}) defined as the ratio of the volumetric flow rate Q (cm^3 min^{-1}) to the cross-sectional area of the bed A (cm^2).

The parameters K_{AB} and N_0 can be calculated from the linear plot of ln (Ce/C_0) against t [10].

2.2.3 Yoon-Nelson model

The Yoon-Nelson model stands out for its straightforward equation notation in contrast to other mathematical models that have been examined. Its fundamental assumption is that the reduction in the rate of adsorption for an adsorbate molecule is directly proportional to the adsorption of the adsorbate and the breakthrough of the adsorbent bed [11]. It assumes that the rate of decay in the probability of adsorption for the adsorbate molecule is proportional to the adsorption of the adsorbate probability and the probability of breakthrough of the adsorbate on the adsorbent [10], and is applicable for a single component system [16]. The nonlinear equation that describes it is the following:

$$\frac{C_e}{C_0} = \frac{e^{k_{YN} * (t-\tau)}}{1 + e^{k_{YN} * (t-\tau)}} \tag{18}$$

The linearized Yoon-Nelson model for a single component system can be expressed as [11]:

$$Ln\frac{C_e}{C_0 - C_e} = K_{YN}t - \tau K_{YN} \tag{19}$$

Where: K_{YN} is the rate constant (min^{-1}) and τ is the time required for 50% adsorbate breakthrough (min).

A linear plot of ln [$Ce/(C_0\text{-}Ce)$] against t determined the values of K_{YN} and τ from the slope and intercept, respectively [8, 10, 16].

2.2.4 The bed-depth/service time (BDST) model

Hutchins in 1973 linearized the model introduced in 1920 by Bohart and Adams. This linearization led to the development of the widely adopted and highly recommended BDST design model. Many authors have since endorsed it as the most straightforward and fastest means of predicting adsorbent performance [17].

The BDST model is employed to describe the characteristics of a column with a consistent bed height. It includes an equation for the duration of operation concerning the deposit depth, which enables the construction of a graph illustrating the relationship between the operation time (t) and the ratio C_t/C_0. This graphical representation facilitates the determination of the dynamic parameters of the BDST model [8, 11].

It is expressed as follows:

$$\mathrm{t} = \frac{N_0}{C_0 F} Z - \frac{1}{K_a C_0} Ln\left(\frac{C_0}{C_t} - 1\right) \tag{20}$$

where: N_0 is the weight of ions adsorbed per unit weight of adsorbate; C_0 is the influent concentration (mg L^{-1}); C_t is ions concentration at breakthrough point (mg L^{-1}); Z is the bed depth (cm); F is the linear velocity (cm h^{-1}); and K_a is the rate constant (L mg^{-1} h^{-1}) [12, 17].

Hutchins expressed the Bohart-Adams equation as a linear equation [12, 13, 17, 18]:

$$t = aZ - b \tag{21}$$

Where:

$$a = \frac{N_0}{C_0 F} \tag{22}$$

$$b = \frac{1}{K_a C_0} Ln\left(\frac{C_0}{C_t} - 1\right) \tag{23}$$

Where: The slope a corresponds to Eq. (22) and the intercept b to Eq. 23. [3, 10, 12, 14, 17, 19–22].

2.2.5 Wolborska model

The Wolborska Model, introduced by Wolborska in 1989, presents an equation that examines the mass transfer of diffusion in breakthrough curves within a low concentration range. The model's equation is depicted as follows:

$$ln\left(\frac{C_t}{C_0}\right) = \frac{\beta C_0 t}{N_0} - \frac{\beta Z}{U_0} \tag{24}$$

Where β is the external mass transfer coefficient min^{-1}. N_0 is the adsorption capacity of the bed per unit volume (mg L^{-1}).

The values of the parameters of the Wolborska model can be determined from the Plot of $ln\left(\frac{C_t}{C_0}\right)$ versus time [23, 24].

2.2.6 Yan et al. model

It overcame the constraint of Thomas's model related to the outlet concentration specifically at time (t = 0).

$$ln\left(\frac{C_t}{C_0 - C_t}\right) = \frac{K_Y C_0}{Q} ln\left(\frac{Q^2}{K_Y q_Y m}\right) + \frac{K_Y C_0}{Q} lnt \tag{25}$$

The values of model parameters are evaluated from the linearized plot between $\ln\left(\frac{C_t}{C_0 - C_t}\right)$ and ln t [1, 24].where K_Y mL mg^{-1} min^{-1}, q_Y mg g^{-1}.

2.2.7 Modified dose-model (MDR)

This model has found widespread application in pharmacology, serving to characterize various processes. Presently, it is applied in delineating column biosorption processes, valued for its precision in depicting the full breakthrough curve while reducing errors, particularly those stemming from the Thomas model, especially in scenarios involving low or high removal times [24]. This model is expressed by:

$$ln\left(\frac{C_t}{C_0 - C_t}\right) = a\ ln\,(C_0 Q_t) - a\ ln\,(q_{mdr} m) \tag{26}$$

By plotting $ln\left(\frac{C_t}{C_0 - C_t}\right)$ against time t, we extract the model parameters, a and q_{mdr}, from the slope and the intercept, respectively. Here, a denotes the model constant, while q_{mdr} signifies the maximum concentration of solute in the solid phase (mg g^{-1}) [24].

The linearized plots of $ln\left(\frac{C_t}{C_0 - C_t}\right)$ versus $\ln(C_0 Q_t)$

2.2.8 Clark model

It coupled mass transfer with the Freundlich isotherm. This model is expressed by:

$$lnA - rt = ln\left(\left(\frac{C_t}{C_0}\right)^{1-n} - 1\right) \tag{27}$$

Values of n required for the above model are acquired from the data of Freundlich isotherm. Linearized plots of $ln\left(\left(\frac{C_t}{C_0}\right)^{1-n} - 1\right)$ versus t [24].

Table 1 presents the models, their equation in linear form, and the main characteristics of the models.

Models	Equation linear form	Main characteristics of the model	References
Thomas Model	$Ln\frac{C_e}{C_0} = K_{AB}C_0 t - K_{AB}N_0\left(\frac{Z}{U_0}\right)$	kinetic model of Pseudo-second order.	[10, 11, 13, 16]
Bohart and Adams Model	$Ln\left[\frac{C_0}{C_e} - 1\right] = \frac{K_{Th}q_e m}{Q} - K_{Th}C_0 t$	Assumed direct proportionality between adsorption rate and adsorbate concentration.	[10, 11]
Yoon-Nelson model	$Ln\frac{C_e}{C_0 - C_e} = K_{YN}t - \tau K_{YN}$	The decrease in adsorption rate is linked to both adsorbate adsorption and breakthrough in the adsorbent bed.	[8, 10, 16]
The bed-depth/service time (BDST) model	$t = az - b$ $a = \frac{N_0}{C_0 F}$ $b = \frac{1}{K_a C_0} Ln\left(\frac{C_0}{C_t} - 1\right)$	Linearized the model by Bohart and Adams	[3, 10, 12, 14, 17, 19–22]

Models	Equation linear form	Main characteristics of the model	References
Wolborska Model	$ln\left(\frac{C_t}{C_0}\right)=\frac{\beta C_0 t}{N_0}-\frac{\beta Z}{U_0}$	Mass transfer of diffusion in breakthrough curves within a low concentration range.	[23, 24]
Yan et al. Model	$ln\left(\frac{C_t}{C_0-C_t}\right)=\frac{K_Y C_0}{Q}\,ln\left(\frac{Q^2}{K_Y q_Y m}\right)+\frac{K_Y C_0}{Q}\,ln\,t$	It addressed the limitation in Thomas's model concerning initial outlet concentration at time (t = 0).	[1, 24]
Modified Dose–Response Model (MDR)	$ln\left(\frac{C_t}{C_0-C_t}\right)=a\,ln\,(C_0 Q_t)-a\,ln\,(q_{mdr} m)$	Widespread application in pharmacology	[24]
Clark Model	$ln\,A-rt=ln\left(\left(\frac{C_t}{C_0}\right)^{1-n}-1\right)$	It coupled mass transfer with the Freundlich isotherm	[24]

Table 1.
Adsorption models for column study.

2.3 Evaluation criteria

To evaluate and compare the experimental and calculated results of C_e/C_o for the dynamic adsorption of ions in a packed bed column system, applied Mean Relative Error (MRE) and Normalized Relative Mean Square Error (NRMSE) statistical criteria. These criteria are described as follows [25]:

$$\mathrm{MRE}=\frac{\sum_{i=1}^{n}\left[\left(\frac{C_e}{C_0}\right)_{cal}-\left(\frac{C_e}{C_0}\right)_{exp}\right]}{n\left(\frac{C_e}{C_0}\right)_{exp}} \qquad (28)$$

Where:
cal is calculated and exp. obtained experimentally.

$$NRMSE=\left[\frac{\sum_{i=1}^{n}\left[\left(\frac{C_e}{C_0}\right)_{cal}-\left(\frac{C_e}{C_0}\right)_{exp}\right]^2}{n\left(\frac{C_e}{C_0}\right)_{exp}}\right]^{0.5}*100 \qquad (29)$$

The performance of the numerical model is poor when the NRMSE >30%, fair if the 20% < NRMSE <30%, good if the 10% < NRMSE <20% and excellent if the 0 < NRMSE <10%. In general, the performance of numerical model is acceptable if the NRMSE <30% [25]. For better assessment, the linear regression was fitted between $(C_e/C_o)_{exp}$ and $(C_e/C_o)_{cal}$ values by the following equation:

$$\left(\frac{C_e}{C_0}\right)_{exp}=m\left(\frac{C_e}{C_0}\right)_{cal}+n \qquad (30)$$

Where: m is the slope of line and n is the distance from the origin.

The value of n is considered zero when it is not significant at 5% level [25].

2.4 Thermodynamic study

Thermodynamic parameters serve to assess the orientation and feasibility of the adsorptive reaction. The conjectures regarding the adsorption mechanism rely on the alterations in Gibbs free energy (ΔG°), the change in standard enthalpy (ΔH°), and the change in standard entropy (ΔS°) [26]. These thermodynamic parameters are calculated using the following equations:

$$K_c = \frac{C_A}{C_e} \tag{31}$$

$$\Delta G^0 = -RT\ ln\, K_c \tag{32}$$

$$\Delta G^0 = \Delta H^0 - T\Delta S^0 \tag{33}$$

$$lnK = \frac{\Delta S^0}{R} - \frac{\Delta H^0}{RT} \tag{34}$$

$$\Delta G^0 = \Delta H^0 - T\Delta S^0 \tag{35}$$

Where: Kc is the adsorption affinity and can be determined by the ratio between the amount adsorbed at equilibrium (C_A mg g^{-1}) and the equilibrium concentration (C_e mg L^{-1}). R is the universal gas constant (8.314 J mol K^{-1}), T is the absolute temperature of the solution (°K). The values of ΔH° and ΔS° are derived from the slope and intercept, respectively, by graphing the Van't Hoff line on the abscissa axis with 1/T against the ordinate axis with ln kc. Alternatively, one can calculate these parameters by graphing ΔG° against T, where the slope of the resulting straight line corresponds to ΔH°, and the intercept represents ΔS°. Alternatively, these parameters can also be calculated by plotting on the abscissa axis with T versus the ordinate axis with ΔG°, whereby the slope of the resulting straight line corresponds to ΔH°, and the intercept represents ΔS° [6, 26–28].

Negative values in ΔG° indicate a spontaneous character of the adsorption process. As the values become more negative with increasing temperature, they indicate that the increase in temperature favors the adsorption process. If negative values are given in ΔH°, this indicates that the process is exothermic, while positive values indicate an endothermic process. When positive values are given in ΔS°, they indicate an increase in randomness at the liquid-solid interface of the system during the adsorption process. If the absolute change in Gibbs free energy is between −20 and 0 KJ mol^{-1}, it is generally physical adsorption, whereas for chemical adsorption, it is in the range of −80 to −400 KJ mol^{-1} [26, 28].

2.4.1 Estimation of activation energy (E_a)

The magnitude of the activation energy provides insight into the nature of the adsorption process, which generally falls into two main categories: physical and chemical adsorption. In activated chemical adsorption, the rate varies with temperature according to a finite activation energy (8.4–83.7 kJ mol^{-1}) in the Arrhenius equation. In the case of non-activated chemical adsorption, the activation energy is close to zero [26, 28]. The activation energy was calculated using the Arrhenius equation:

$$K_c = K_0 Exp\left(-\frac{E_a}{RT}\right) \tag{36}$$

Where:

k_0 is the temperature-independent rate constant (g mg^{-1} min^{-1}),

E_a is the apparent activation energy of the adsorption reaction (kJ mol^{-1}). The linear form of the equation is expressed as:

$$Ln\ K_c = -\frac{E_a}{R}\frac{1}{T} + LnK_0 \tag{37}$$

By plotting ln K_c against 1/T, we will obtain a straight line with a slope equal to $-E_a/R$ [26, 28].

2.5 Scale-up design

Given the resemblances in empty bed contact time (EBCT) and hydrodynamic characteristics between laboratory-scale and pilot/full-scale column systems, it is viable to employ data from the laboratory scale for calculating parameters and modeling the efficiency of larger column systems using vertical scaling techniques. To achieve comparable hydrodynamic properties between laboratory-scale and large-scale column systems, this approach uses the filtration rate (FR) and empty bed contact time (EBCT) values obtained from a laboratory-scale column system as the basic parameters for the design of the larger column system. These relationships are defined by the following equations [15]:

$$FR = \frac{Q_L}{A_L} \tag{38}$$

$$A_D = \frac{Q}{FR} \tag{39}$$

$$\tau = \frac{V_L}{Q} \tag{40}$$

$$H_{BD} = FR * \tau \tag{41}$$

$$M_D = V_D * A_D * H_{BD} \tag{42}$$

$$B_D = \frac{M_{AC}}{M_{CR}} \tag{43}$$

$$BV_D = B_D * Q_D \tag{44}$$

Where: FR is the filtration rate (cm min^{-1}), A_D area of the design column (cm^2), EBCT of the lab-scale column (τ, min), H_{BD} bed height of the design column (cm), M_D mass of the adsorbent required in the design column (kg), B_D breakthrough time of the design column (min), BV_D volume treated before breakthrough (m^3), Q_L is flow rate in the lab-scale (cm^3 min^{-1}), A_L denotes cross-sectional area in the lab-scale (cm^2), V_L volume of lab-scale column (cm^3), V_D volume of design column (cm^3), M_{AC} is amount of adsorbent consumed in the design column (kg), M_{CR} is adsorbate consumption rate of the design column (kg d^{-1}), and Q_D is flow rate in the design column (m^3 d^{-1}). Since the ratio of the column diameter to the particle diameter is more than twenty, wall effects can be assumed negligible [29].

3. Conclusions

To optimize outcomes when implementing an adsorption-based treatment system, starting with batch experiments is key. These experiments are straightforward and cost-effective, offering preliminary insights into the adsorption traits of a particular material. However, they often fall short in fully establishing an effective treatment process. To achieve this, conducting continuous experiments becomes essential for a more comprehensive understanding of a material's adsorption behaviors in a continuous setup. Detailed mathematical models designed for this purpose have been extensively developed to elucidate how a material interacts with specific contaminants or a combination thereof. Once results from a laboratory-scale column are obtained, it is advisable to proceed by scaling up and testing using a pilot or semi-pilot plant-scale model to ensure the reliability of the findings.

Verifying the similarity between results derived from mathematical models and those obtained experimentally is crucial. Using evaluation criteria is necessary to ensure a level of certainty regarding the statistical significance and resemblance of the experimental results.

Acknowledgements

Advanced Materials Research Center (CIMAV) and Faculty of Animal Science and Ecology, Autonomous University of Chihuahua.

Conflict of interest

The authors declare no conflict of interest.

Author details

Andrés A. Abin-Bazaine[1*], Mario A. Olmos-Marquez[2] and Alfredo Campos-Trujillo[1]

1 Advanced Materials Research Center (CIMAV), Complejo Industrial Chihuahua, Chihuahua, México

2 Faculty of Animal Science and Ecology, Autonomous University of Chihuahua, Chihuahua, México

*Address all correspondence to: abinsdreamkennel@gmail.com

References

[1] Patel H. Fixed-bed column adsorption study: A comprehensive review. Applied Water Science [Internet]. 2019;**9**(3):1-17. DOI: 10.1007/s13201-019-0927-7

[2] Thirunavukkarasu A, Nithya R, Sivashankar R. Continuous fixed-bed biosorption process: A review. Chemical Engineering Journal Advances. 2021;**8**:100188

[3] Oguz E, Ersoy M. Removal of Cu^{2+} from aqueous solution by adsorption in a fixed bed column and neural network modelling. Chemical Engineering Journal. 2010;**164**(1):56-62

[4] Arim AL, Neves K, Quina MJ, Gando-Ferreira LM. Experimental and mathematical modelling of Cr(III) sorption in fixed-bed column using modified pine bark. Journal of Cleaner Production. 2018;**183**:272-281

[5] Al-Saydeh SA, El-Naas MH, Zaidi SJ. Copper removal from industrial wastewater: A comprehensive review. Journal of Industrial and Engineering Chemistry. 2017;**56**(July):35-44

[6] Espina, de Franco MA, Bonfante de Carvalho C, Marques Bonetto M, de Pelegrini SR, Amaral FL. Diclofenac removal from water by adsorption using activated carbon in batch mode and fixed-bed column: Isotherms, thermodynamic study and breakthrough curves modeling. Journal of Cleaner Production. 2018;**181**:145-154

[7] Rosales E, Meijide J, Pazos M, Sanromán MA. Challenges and recent advances in biochar as low-cost biosorbent: From batch assays to continuous-flow systems. Bioresource Technology. 2017;**246**:176-192

[8] Bai S, Li J, Ding W, Chen S, Ya R. Removal of boron by a modified resin in fixed bed column: Breakthrough curve analysis using dynamic adsorption models and artificial neural network model. Chemosphere. 2022;**296**:134021

[9] Meshram S, Dharmadhikari S, Thakur RS, Soni AB, Thakur C. Fixed-bed adsorption of lead from battery recycling unit wastewater-optimization using box-Behnken method. Journal of Hazardous Materials Advances. 2023;**10**:100297

[10] Chen S, Yue Q, Gao B, Li Q, Xu X, Fu K. Adsorption of hexavalent chromium from aqueous solution by modified corn stalk: A fixed-bed column study. Bioresource Technology. 2012;**113**:114-120

[11] Szostak K, Hodacka G, Długosz O, Pulit-Prociak J, Banach M. Sorption of mercury in batch and fixed-bed column system on hydrochar obtained from apple pomace. Processes. 2022;**10**(10):2114

[12] Ji F, Li C, Xu J, Liu P. Dynamic adsorption of Cu(II) from aqueous solution by zeolite/cellulose acetate blend fiber in fixed-bed. Colloids and Surfaces A: Physicochemical and Engineering Aspects. 2013;**434**:88-94

[13] Du Z, Zheng T, Wang P. Experimental and modelling studies on fixed bed adsorption for Cu(II) removal from aqueous solution by carboxyl modified jute fiber. Powder Technology. 2018;**338**(Ii):952-959

[14] Oguz E. Fixed-bed column studies on the removal of Fe^{3+} and neural network modelling. Arabian Journal of Chemistry. 2017;**10**(3):313-320

[15] Jung KW, Jeong TU, Choi JW, Ahn KH, Lee SH. Adsorption of phosphate from aqueous solution using electrochemically modified biochar calcium-alginate beads: Batch and fixed-bed column performance. Bioresource Technology. 2017;**244**(July):23-32

[16] Lim AP, Aris AZ. Continuous fixed-bed column study and adsorption modeling: Removal of cadmium (II) and lead (II) ions in aqueous solution by dead calcareous skeletons. Biochemical Engineering Journal. 2014;**87**:50-61

[17] Cardona Y, Korili SA, Gil A. Use of response surface methodology to optimize triclosan adsorption on alumina pillared clays in a fixed-bed column for applications in solid-phase extraction. Applied Clay Science. 2023;**235**:106879

[18] Talat M, Mohan S, Dixit V, Singh DK, Hasan SH, Srivastava ON. Effective removal of fluoride from water by coconut husk activated carbon in fixed bed column: Experimental and breakthrough curves analysis. Groundwater for Sustainable Development. 2017;7:48-55

[19] Hernandez-Eudave MT, Bonilla-Petriciolet A, Moreno-Virgen MR, Rojas-Mayorga CK, Tovar-Gómez R. Design analysis of fixed-bed synergic adsorption of heavy metals and acid blue 25 on activated carbon. Desalination Water Treatment. 2016;**57**(21):9824-9836

[20] Han R, Wang Y, Zhao X, Wang Y, Xie F, Cheng J, et al. Adsorption of methylene blue by phoenix tree leaf powder in a fixed-bed column: Experiments and prediction of breakthrough curves. Desalination. 2009;**245**(1–3):284-297

[21] Kavand M, Fakoor E, Mahzoon S, Soleimani M. An improved film–pore–surface diffusion model in the fixed-bed column adsorption for heavy metal ions: Single and multi-component systems. Process Safety and Environmental Protection. 2018;**113**:330-342

[22] Abdolali A, Ngo HH, Guo W, Zhou JL, Zhang J, Liang S, et al. Application of a breakthrough biosorbent for removing heavy metals from synthetic and real wastewaters in a lab-scale continuous fixed-bed column. Bioresource Technology. 2017;**229**:78-87

[23] Awad AS, Hudaib B, Omar W. Modeling date palm trunk fibers (DPTF) packed bed adsorption performances for cadmium removal from aqueous wastewater. Fluid Dynamics and Materials Processing. 2023;**19**(6): 1535-1549

[24] Banerjee M, Bar N, Das SK. Cu(II) removal from aqueous solution using the walnut shell: Adsorption study, regeneration study, plant scale-up design, economic feasibility, statistical, and GA-ANN modeling. International Journal of Environmental Research. 2021;**15**(5):875-891

[25] Amiri MJ, Roohi R, Gil A. Numerical simulation of Cd(II) removal by ostrich bone ash supported nanoscale zero-valent iron in a fixed-bed column system: Utilization of unsteady advection-dispersion-adsorption equation. Journal of Water Process Engineering. 2018;**25**(May):1-14

[26] Avelino Abin-Bazaine A, Ríos GSA-DL, Rodríguez-Vázquez LM, Santellano-Estrada E, Rodríguez-Piñeros S, Cortés-Palacios L. Copper removal by acid-conditioned zeolite, part II: Kinetics, and thermodynamic studies. Journal of Environment and Earth Science. 2019;**9**: 39-50. Available from: www.iiste.org

[27] de Franco MAE, de Carvalho CB, Bonetto MM, de Soares RP, Féris LA.

Removal of amoxicillin from water by adsorption onto activated carbon in batch process and fixed bed column: Kinetics, isotherms, experimental design and breakthrough curves modelling. Journal of Cleaner Production [Internet]. 2017;**161**:947-956. Available from: https://linkinghub.elsevier.com/retrieve/pii/S0959652617311496

[28] Avelino Abin-Bazaine A, Ríos GSA-DL, Miguel Rodríguez-Vázquez L, Santellano-Estrada E, Rodríguez-Piñeros S, Cortés-Palacios L. Copper removal by acid-conditioned zeolite, part II: Kinetics, and thermodynamic studies. Journal of Environment and Earth Science [Internet]. 2019;**9**(3): 39-50. Available from: www.iiste.org

[29] Yan L, Huang Y, Cui J, Jing C. Simultaneous As(III) and Cd removal from copper smelting wastewater using granular TiO_2 columns. Water Research. 2015;**68**(Iii):572-579

Chapter 4

Thermodynamic Properties of Moisture Sorption and Glass Transition of Coconut (*Cocos Nucifera* L.) Powder Fortified with Physiologically Active Components

Juan Carlos Lucas-Aguirre,
German Antonio Giraldo-Giraldo and Misael Cortés-Rodríguez

Abstract

The purpose of this work is to determine the thermodynamic properties of coconut powder fortified with physiologically active components, obtained through spray drying ($CP + FAC$), using the data obtained to construct the sorption isotherms at three temperatures (15, 25, and 35°C), where the properties calculated were net integral isosteric sorption heat (Q_{st}), differential sorption entropy (ΔS), spreading pressure (Φ), enthalpy (q_{eq}), and entropy (ΔS_{eq}); whereas determining the glass transition temperature (T_g) and critical storage conditions, such as water activity (a_{wc}), and moisture content (X_{wc}) in the $CP + FAC$. Q_{st} increased to a maximum value between the X_w intervals of the monolayer and then diminished with increased moisture content (X_w). The Φ increased with the a_w increase and diminished with increased temperature, more notably at 25 and 35°C. The ΔS diminished with the increase of X_w, representing an increase in the mobility restriction of the water molecules as the available sites saturate and higher-energy sites are used, which is why water is less available to participate in the deterioration reactions. The q_{eq} diminished with X_w, reaching the maximum value at the lowest X_w; this value is an indication of the greater water-$CP + FAC$ interaction that means the sorption binding sites are stronger. The ΔS_{eq} went from a negative value to an increased X_w, which is associated with greater mobility of the water molecules, favoring the formation of multilayers. The glass transition temperature of the $CP + FAC$ at different a_w had an inverse relation, where T_g diminished with increasing a_w, going to a_w: 0.112: T_g: 69.79°C at a_w values of 0.900: T_g: −39.0°C, evidencing the plasticizer effect of water, where the Gordon-Taylor model is a reliable predictor of the glass transition temperatures of the $CP + FAC$. The values of the parameters calculated through nonlinear regression were T_{gs} = 391.67 K and k = 0.753, with R^2 = 0.977.

IntechOpen

The critical storage conditions of the *CP + FAC*, at an ambient temperature of 35°C, were a_{wc} = 0.46 and X_{wc} = 3.8% (*db*).

Keywords: thermodynamic sorption properties, differential scanning calorimetry, glass transition temperature, activity and water critical content, stability

1. Introduction

The study of water-solid interactions in low-moisture food systems provides useful information that integrates the composition and structural characteristics with the physical stability and functional properties of food materials. In addition, the thermodynamic analysis of water sorption in dry foods at different temperatures has stirred interest because it provides a complete interpretation of the sorption mechanism based on the energy changes in the bonds between the adsorbate and adsorbent in dry foods, such as coconut (*Cocos Nucifera* L.) powder fortified with physiologically active components (*CP + FAC*) [1, 2].

Linked to the phenomena of moisture sorption, the analysis of glass transition temperature (T_g) permits a better understanding of the quality, stability, and safety properties of food systems in function of the relation between the water content and the matrix structure, thus, providing an integrated approach of the role of water in foods. Glass transition, understood as the change from a highly viscous amorphous glass state to a liquid or gummy state, can be reached through increased temperature with constant moisture content or through increased water content at constant temperature. Glass transition can be classified approximately as a second-order phase change accompanied by thermodynamic changes in the enthalpy, changes in the dielectric properties, and mechanical changes, such as increased free volume, diminished viscosity, increased thermal expansion, and exponential increase of the molecular mobility. Foods in an amorphous vitreous state exist in a metastable condition and remain stable for long periods (months to years); however, once they pass to the gummy state, all the rates of processes of quality loss dependent on time increase, and the shelf life diminishes to weeks, days, or even hours. Identifying the temperatures and critical moisture contents, at which amorphous solids experience evident rheological changes, contributes to the correct definition of the most adequate storage conditions where time-dependent changes, deterioration reactions, and structural transformations, such as adherence, collapse, stickiness, caking, and crystallization of solutes from powdered foods would be limited and/or prevented [1–8].

As temperature increases above T_g, various changes are noted, such as increased free volume, diminished viscosity, and increased thermal expansion. The most important changes that affect food behavior are related with the exponential increase of molecular mobility and diminished viscosity [5, 6]. Because low glass transition temperature is a result of the presence of low molecular weight substances such as reducing sugars or acids (for example citric, fumaric, or malic acids), powdered fruits are very vulnerable to deterioration changes during processing and storage. The widely applied approach allows obtaining powders with higher T_g values in addition to high molecular weight polymers such as starches, arabic gum, or maltodextrin; and the influence of moisture on T_g may be examined easily via thermal analyses through differential scanning calorimetry (*DSC*) in samples that have been balanced for different water activities (a_w), and thereby, for different moisture contents [7, 8].

Another consistent criterion to predict the storage stability and shelf life of dehydrated products is based on determining the thermodynamic properties of water vapor sorption in the function of temperature (isosteric sorption heat, differential entropy, integral entropy, integral enthalpy, and spreading pressure); for which, sorption isotherms, which describe the relation between a_w and the equilibrium moisture content of food at constant temperature are fundamental to comprehend the adsorbate and adsorbent relations, guarantee that water is not available, and to limit the chemical reactions and deterioration of physical–chemical and microbiological characteristics of powdered foods [1, 3, 6, 9].

The thermodynamic properties of foods relate the water concentration in foods with its partial pressure, which is crucial in the analysis of heat and mass transport phenomena during dehydration, and permits determining the final point at which food should dehydrate to achieve a stable product with optimal moisture content and the minimum amount of energy required to eliminate a given amount water from the food, providing information on the microstructure associated with the water-food interface [6, 10–14].

Differential sorption heat, sometimes called isosteric sorption heat (Q_{st}), is used as an indicator of the state of the water adsorbed by the solid particles, which is a measure of the physical, chemical, and microbial stability of the biological materials stored. Knowing the differential sorption heat is important when designing equipment for dehydration processes. This is because the vaporization heat of the water sipped can increase to values above the vaporization heat of pure water as the foods dehydrate to low moisture levels [2, 10, 12].

The net isosteric heat (q_{st}) is defined as the total sorption heat in foods less the vaporization heat of water, at the system's temperature. Adsorption heat is the measure of the energy released in the sorption, and desorption heat requires energy to break the intermolecular forces between the water vapor molecules and the surface of the adsorbent. Thereby, sorption heat is considered indicative of the intermolecular attraction forces between the sorption sites and water vapor [4, 10, 12–15].

The differential entropy (ΔS) of a material is proportional to the number of sorption sites available at a specific energy level, associated with the molecular ordering system, if $\Delta S < 0$, it describes a structured system [2, 10, 12–15].

Two other parameters used to explain the modes of moisture adsorption by foods are integral entropy (ΔS_{eq}) and integral enthalpy (q_{eq}). Integral entropy describes the degree of disorder and randomness of the water molecule movement and has been used to explain modes of moisture sorption by biological materials. The net integral enthalpy, or net equilibrium sorption heat, is an integral molar amount and is calculated similarly to the differential sorption heat (isosteric sorption heat) but at constant spreading pressure instead of constant moisture content [1, 4, 10, 12].

Spreading pressure (Φ) and its dependence on a_w and water temperature is a useful tool to interpret kinetic sorption data, with the driving force responsible for diffusion in porous solids and/or as the adsorption-free surface energy, possibly considered as the difference in surface tension between the sorption sites discovered in the solid and sites with adsorbed molecules [10, 12].

Food safety and spoilage are the principal concerns of food manufacturers and consumers. The preservation achieved by dehydrating powdered foods requires information on the thermodynamic properties related with water activity and vitrification phenomena. Efforts have been limited to establishing a link between a_w and T_g for specific food products, as in *CP* + *FAC*. This work sought to calculate and evaluate the thermodynamic properties of water adsorption and determine the glass transition

Model	Mathematical expression
GAB [19]	$X_w = m_o CKa_w/[(1 - Ka_w)(1 - Ka_w + CKa_w)]$
BET [20]	$X_w = m_0 Ca_w/[(1 - a_w)(1 - a_w + Ca_w)]$
OSWIN [21]	$X_w = A[a_w/\left(1 - a_w\right)]^B$
SMITH [22]	$X_w = c_1 - c_2 \ln(1 - a_w)$
HALSEY [23]	$X_w = \left[-A/\left(T\ln a_w\right)\right]^{1/B}$
HENDERSON [24]	$X_w = \left[-\ln(1 - a_w)/A\right]^{1/B}$
CHUNG and PFOST [25]	$X_w = a + b\ln(-\ln a_w)$
PELEG [26]	$X_w = m_1 {a_w}^{n1} + m_2 {a_w}^{n2}$
CAURIE [27]	$X_w = \exp(a + ba_w)$

Where: moisture content (X_w) *in dry base;* m_o*: humidity of the monolayer;* a, b, c_1, c_2, m_1, m_2, n_1, n_2, A, B, C, K *are adjustment parameters of each model, which were estimated with the Polymath® program version 6.0;* T: *Temperature K.*

Table 1.
Mathematical models used to fit the water sorption isotherms in CP + FAC.

temperature of *CP + FAC* at different concentrations of solids, as well as the critical storage conditions.

2. Materials and method

To obtain the *CP + FAC* and the construction of the moisture sorption isotherms, the study followed the methodologies reported by Lucas et al. [16–18] respectively.

To fit the moisture sorption isotherms in the *CP + FAC*, nine mathematical models were used according to the equations by *GAB* [19], *BET* [20], Oswin [21], Smith [22], Halsey [23], Henderson [24], Chung and Pfost [25], Peleg [26], and Caurie [27] (**Table 1**), and the goodness-of-fit of the models was determined by using the mean square error (E_{RSM}) and the maximum adjusted R^2.

2.1 Determination of the thermodynamic properties

From the data obtained in determining the moisture sorption isotherms of *CP + FAC*, these were calculated with the following methodologies:

2.1.1 Isosteric sorption heat ($kJ \cdot mol^{-1}$)

The net isosteric heat (q_{st}) and isosteric sorption heat (Q_{st}), on the basis of thermodynamic principles, were determined from Eqs. (1) and (2) as follows:

$$q_{st} = -R\left[\frac{d(\ln a_w)}{d(1/T)}\right]_{EMC} \tag{1}$$

$$Q_{st} = q_{st} + \lambda_{vap} \tag{2}$$

Where λ_{vap}: latent vaporization heat of pure water ($kJ \cdot mol^{-1}$); T (K): absolute temperature; X_{eq}: equilibrium moisture content ($kg \cdot kg^{-1}$ db), and R: ($kJ \cdot mol^{-1} \cdot K^{-1}$) ideal gas constant.

This was calculated from the equation by Clausius-Clapeyron applied to the system and the pure water with the following assumptions: (1) the vaporization heat of pure water and excess sorption heat do not change with temperature and (2) the system's moisture content remains constant. When again plotting the sorption isotherm in the form ln (a_w) versus 1/T, for a specific moisture content, the slope of the regression line provides a measure of the net isosteric sorption heat, q_{st}, obtaining the sorption heat Q_{st} by applying Eq. (2). This procedure assumes that q_{st} is invariant with temperature, and the application of the method requires measuring sorption isotherms at more than two temperatures [1, 2, 10, 14, 28–30].

2.1.2 Spreading pressure (Φ) ($J \cdot m^{-2}$)

This is also known as surface potential, which represents the free surface energy for sorption and may be considered as the difference in surface tension between the active adsorption sites in the solid and water molecules. The value Φ is not determined experimentally, but it can be obtained through an analytical procedure related with Eq. (3) [10, 12, 29].

$$\Phi = \frac{K_B\, T}{A_m} \int_0^{a_w} \frac{\theta}{a_w}\, d(a_w) \tag{3}$$

Where $\theta = EMC/x_o$ (x_o: moisture in the monolayer).

Applying the Dent model [31], where it is expressed by integrating the spreading pressure of water adsorbed in terms of surface area per sorption site, or area per water molecule in each sorption site, that is, coverage of the complete monolayer as follows:

$$\Phi = \frac{K_B\, T}{A_m} \ln\left[\frac{1 + b_0\, a_w - b\, a_w}{1 - b\, a_w}\right] \tag{4}$$

Where: K_B: Boltzmann's constant ($1.380 \times 10^{-23}\ J \cdot K^{-1}$); A_m: area of water molecule: ($1.06 \times 10^{-19}\ m^2$); b and b_0 are the constants of Dent's sorption isotherm related with the properties of the water adsorbed (dimensionless), determined through nonlinear regression analysis using the moisture content of the monolayer obtained by applying the *BET* equation to the experimental data of equilibrium moisture.

2.1.3 Differential entropy (ΔS) ($J \cdot mol^{-1} \cdot K^{-1}$)

The differential adsorption entropy was calculated by fitting Eq. (5) to the equilibrium data [10, 28, 29].

$$-\ln(a_w) = \frac{-Q_{st}}{RT} - \frac{\Delta S}{R} \tag{5}$$

When tracing ln (a_w) versus 1/T for constant water content, the value of ΔS is calculated from the intersection ($\Delta S/R$).

2.1.4 Net integral sorption enthalpy (q_{eq}) ($J \cdot mol^{-1} \cdot K^{-1}$)

The net integral enthalpy is considered the difference in surface tension between the bare sorption sites in the solid and sites with adsorbed molecules; it is an integral molar amount and is calculated similarly to the isosteric sorption heat but at constant spreading pressure, where the variation in net integral enthalpy with moisture content indicates the level at which the water/substrate interaction is greater than the interaction of the water molecules (Eq. 6) [10, 12, 29].

$$q_{eq} = -R\left[\frac{d(\ln a_w)}{d(1/T)}\right]_{\Phi} \tag{6}$$

2.1.5 Net integral sorption entropy (ΔS_{eq}) ($J \cdot mol^{-1} \cdot K^{-1}$)

Integral enthalpy is needed to determine the integral entropy associated with the sorption process. Net integral entropy describes the degree of disorder and randomness of the water molecule movement. It also quantifies the mobility of the water molecules adsorbed during sorption. The ΔS_{eq} of a system may be calculated with Eq. (7), considering that the geometric mean of activity ($a_w{}^*$) is used, obtained at constant spreading pressure at various temperatures [10, 12, 28, 29]:

$$\Delta S_{eq} = \frac{-Q_{in}}{T} - R\ln(a_w *) \tag{7}$$

2.2 Glass transition temperature (T_g) in the CP + FAC

A thermal analysis was performed of the *CP + FAC* by using differential scanning calorimetry (*DSC*; T.A. Instrument, Q2000, USA) equipped with a cooling unit up to −90°C (T.A. Instrument, refrigerated Cooling System 90). Samples of *CP + FAC* (5–10 mg) obtained from the balanced isotherms of 15°C and different a_w were analyzed according to the methodology proposed by Grabowski *et al.* [32] (modified), using a cooling ramp of 10°C/min until reaching −90°C, 1-min holding time at −90°C and a heating ramp of 10°C/min until 250°C. The thermograms obtained were examined by using the Universal Analysis 2000 Program (T. A. Instrument, USA), with prior calibration of temperature and fusion heat with indium [1, 5, 6, 33].

2.3 Plasticizer behavior of water

To evaluate and predict the plasticizer effect of water and its influence on the glass transition temperature (T_g), model of the linearized Gordon-Taylor model was used (Eq. 8) [1, 2, 5–8, 33, 34]:

$$T_g = T_{gs} + k\frac{x_w(T_{gw} - T_g)}{(1 - x_w)} \tag{8}$$

Where T_{gw} is the water glass transition temperature: −135.15°C (138.15 K), T_{gs} is the glass transition temperature of the anhydrous solids, x_w is the mass fraction of water, and k is the constant of the Gordon-Taylor model.

2.4 Particle size distribution

Particle sizes were determined as percentiles D_{10}, D_{50}, and D_{90}, using a Mastersizer 3000 (Malvern Instrument Ltd., Worcestershire, UK), with prior dispersion of the samples in 500 mL of distilled water until obtaining a darkening value of $10 \pm 1\%$, considering the size distribution from Mie theory and using the refraction index of 1.52 [35].

2.5 Particle morphology

This was carried out through *SEM* analysis in the *CP* + *FAC*. Micrographs were taken of the *CP* + *FAC* by using scanning electron microscopy (Jeol 5910LV) at 15 Kv, where the samples were deposited on a copper conductive tape and on a sample holder, then coated with gold in a vacuum evaporator (Denton Vacuum, 30 mA, 5 kV, 100 militorr) [33, 36].

The particle size distribution and morphology of the *CP* + *FAC* were carried out before and after storing the powder at 35°C, in N_2 atmosphere (35°C-N_2), packaged in multilayered bags (Alico S.A.), with laminated film thickness (*PET*) = 12 μm, aluminum foil = 8 μm, and polyethylene sealing layer = 100 μm, grammage of 136.54 g · m^{-2}, with water vapor barrier <5 $cm^3 \cdot m^{-2} \cdot$ 24 h · atm), and O_2 barrier <5 $cm^3 \cdot m^{-2}$ · 24 h · atm).

3. Results and discussion

3.1 Water sorption isotherms of the CP + FAC

The water adsorption isotherms of the *CP* + *FAC* at the three temperatures evaluated had typical Type II behavior, according with the *BET* classification [37], with sigmoidal shape, as presented in **Figure 1**, which are characteristics of amorphous materials rich in hydrophilic components. Sigmoid isotherms may be divided generally into the following three regions: Region I corresponds to $a_w < 0.22$, which refers to water adsorption in the monolayer (m_o); Region II represents a_w between 0.22 and 0.73, corresponding to the water adsorption multilayers additional to the m_o; and in Region III, a_w is between 0.73 and 1.0, corresponding to water condensation in the pores of the material followed by the material's solubility [11, 18, 19, 38].

According to Lucas et al. [18], the mathematical models that best fit the behavior of the data of the moisture adsorption isotherms in the *CP* + *FAC*, were the Peleg [26], *GAB* [19], and *BET* [20] models (**Table 2** and **Figure 1**), presenting the lowest E_{RSM} values and highest R^2 values, for all temperatures.

Figure 1 shows the experimental moisture sorption isotherms for the *CP* + *FAC* and those represented by the Peleg model, fitting adequately to the experimental results. The X_w of the monolayer (m_o) shows the amount of water adsorbed strongly on the food surface and is considered the optimal value to ensure the product's stability; for the *CP* + *FAC*, it was observed that in the *GAB* [19] and *BET* [20] models, the moisture of the m_o diminished when temperature increased, showing values between 7.15% and 3.64% at 15 and 25°C for *GAB* and from 2.54% to 2.34% at the same temperatures for *BET*. This behavior indicates that the molecules absorbed gained kinetic energy causing the attractive forces to be loosened, and this permitted some water molecules to separate from their sorption sites, thus, diminishing the

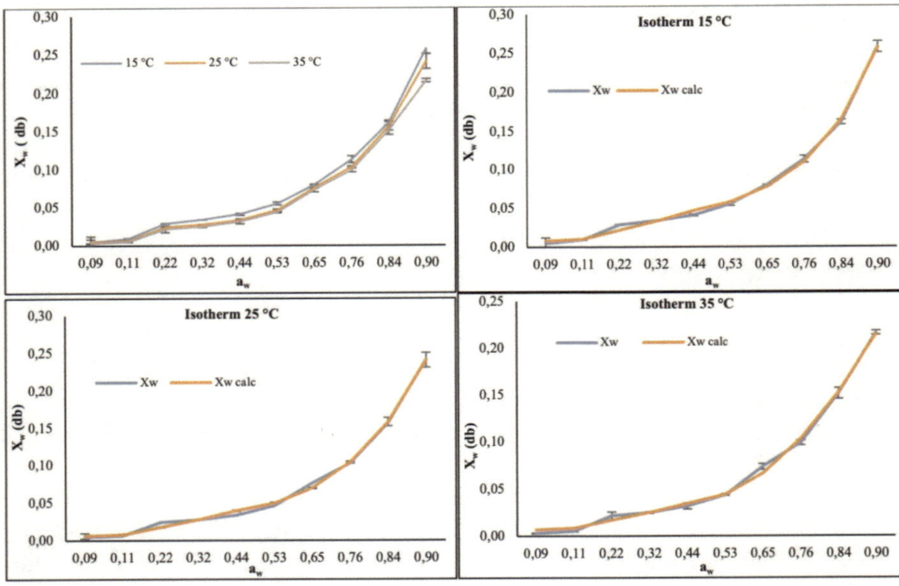

Figure 1.
Water sorption isotherms of experimental and modeled CP + FAC *[26].*

Model	Variable	15°C	25°C	35°C
GAB	C	2.64	4.261	1.811
	m_o	0.0715	0.0573	0.0364
	K	0.921	0.93	0.941
	R^2	0.988	0.994	0.997
	E_{RSM}	0.53	0.415	0.266
BET	C	6.555	5.921	6.365
	m_o	0.0254	0.0254	0.0234
	R^2	0.987	0.994	0.984
	E_{RSM}	0.296	0.294	0.347
PELEG	A	0.467	0.103	0.076
	B	1.308	1.18	1.009
	C	0.118	0.395	0.294
	D	10.668	9.249	6.639
	R^2	0.998	0.997	0.997
	E_{RSM}	0.198	0.199	0.276

Table 2.
Statistical parameters of the goodness of fit of the most precise mathematical models.

equilibrium humidity values. Additionally, moisture of the m_o in the *BET* model was lower than that of the *GAB* model, whereas the energy constant *C* in the BET model was higher than that of the *GAB* model. This behavior has been observed by various authors (**Table 2**). Presumably, the strong adsorbent-adsorbate interactions, which are exothermic, are favored with lower temperature, causing increased parameter C with decreasing temperatures [15, 34].

3.2 Thermodynamic properties of the CP + FAC

3.2.1 Isosteric sorption heat (Q_{st})

This is used to study the magnitude of binding energy, or the availability of polar sites to water vapor and of the physical-chemical and microbiological state of the food, as sorption advances. Isosteric adsorption heat increased to a maximum value, within the moisture content interval of the monolayer (m_o; 0.48%–2.87% [d]) found in the *GAB* and *BET* models of X_w, whose values ranged between 35.718 and 99.261 $kJ \cdot mol^{-1}$ and then diminished with increased moisture content. The maximum value indicates the coverage of the strongest binding sites and the highest water-solid interaction (adsorbate-adsorbent), and water is closely bonded to the material, which corresponds to high interaction energy. Upon continuing the coating of favorable sites and the formation of multiple layers, as shown by diminished Q_{st}, with increased moisture content, this confirms the fact that higher moisture levels diminish the binding force to water (**Figure 2**). The high Q_{st} values at low moisture contents indicate high water-binding energy, which is characteristic of monolayer sorption; this is explained by the fact that most sites with high water binding energies were already occupied but sorption continued on sites with lower water binding energies. The same behavior was observed in sugar cane powder [15], chia seeds [4], cassava flour [12], dried and pulverized edible house cricket, and black soldier fly larvae [30].

The high values at low moisture contents can be explained by the strong hydrogen bonding of water molecules to the food constituting a monolayer of molecules. Therefore, the amount of energy required to remove these water molecules is high. The decreasing trend of Q_{st} suggested that the heat of sorption approaches the heat of vaporization of pure water at higher moisture contents. As the moisture content decreases, and only the monolayer moisture is left, the water molecules become more strongly bound to the surface of the powders and to the sorption sites with higher interaction energies. At the same time, the heat of sorption increases above the heat of vaporization of pure water, making it difficult to remove water from the surface of the powders [39].

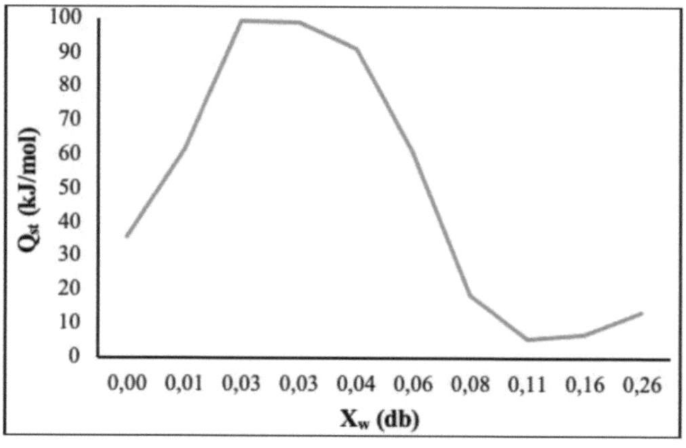

Figure 2.
Isosteric sorption heat of the CP + FAC.

3.2.2 Spreading pressure (Φ) ($J \cdot m^{-2}$)

The value of Φ represents excess free energy on the surface, which indicates increased surface tension in the sorption sites, caused by molecules adsorbed. **Figure 3** shows the behavior of Φ, whose values increase with increased a_w and diminish with increased temperature, most notably at 25 and 35°C [29, 40].

3.2.3 Differential sorption entropy (ΔS) ($J \cdot mol^{-1} \cdot K^{-1}$)

Figure 4 shows the ΔS in function of moisture content at 15, 25, and 35°C, where *CP + FAC* diminished as it gained moisture until reaching a minimum value and then increased as the X_w continued increasing, although—in general—there were no significant differences due to the effect of temperature. Diminished ΔS represents an increase in the mobility restriction of the water molecules as the available sites saturate and sites of greater energy are used; the subsequent increase implies that the water molecules can form multilayers. The ΔS is considered that of maximum stability

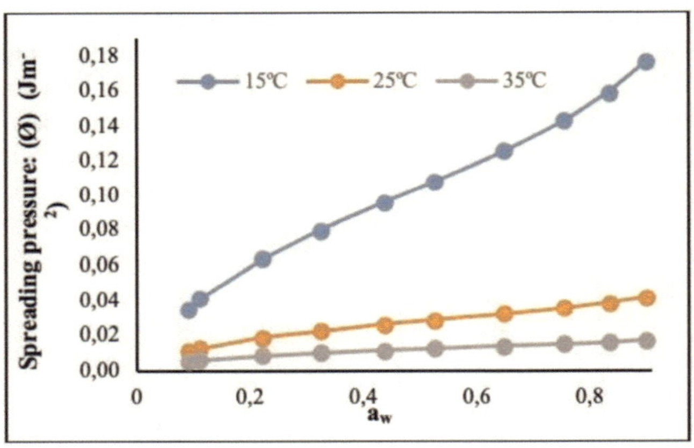

Figure 3.
Spreading pressure of the CP + FAC.

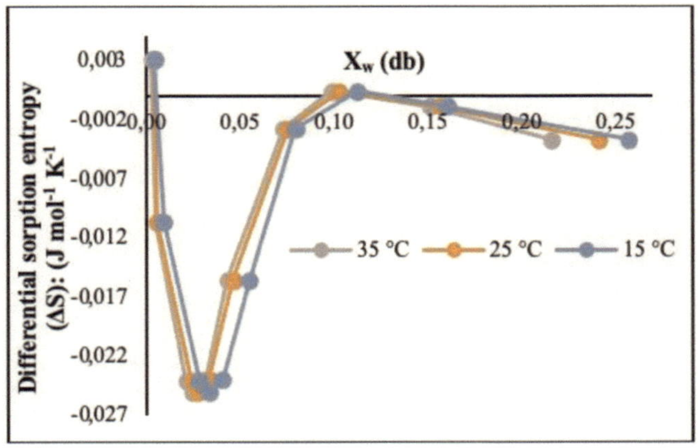

Figure 4.
Differential sorption entropy of the CP + FAC.

because it is where the water molecules achieve a neater disposition within the solid; and strong bonds are produced between the adsorbate and adsorbent, which is why water is less available to participate in deterioration reactions. With very high moisture contents, entropy will approximately be the same as in liquid water because interactions between water and *CP* + *FAC* will be null [1, 4, 15].

Conditions for the maximum stability of *CP* + *FAC* were 2.87 kg H_2O per 100 kg s.s. (a_w = 0.29) at 15°C; 2.44 kg H_2O per 100 kg s.s. (a_w = 0.33) at 25°C; and 2.14 kg H_2O per 100 kg s.s. (a_w = 0.37) at 35°C (**Figure 4**). In addition, integral entropy may be directly related with the disorder of the order of the water molecules sorbed in foods and, hence, it is a useful function to study the effect of the drying method on the product's stability. The same behavior was reported by Velázquez-Gutiérrez *et al.* [4] with chia mucilage, Alpizar-Reyes *et al.* [1] with tamarind seed mucilage, and Ayala-Aponte *et al.* [12] with cassava flour.

Although no differences were noted in ΔS because of the effect of temperature, normally, at higher temperatures, the change of ΔS is greater because the kinetic energy of the molecules that interact during the water exchange process is directly proportional to the temperature. Thereby, the system's ΔS is lesser at low temperatures due to the diminished number of sorption sites and the restricted movement of the water molecules, which makes the adsorption process more favorable thermodynamically at lower temperatures. High ΔS values at low moisture contents are explained by the fact that water was strongly bound, but lower ΔS values may be interpreted as the water activity at which this product is more stable [4, 12, 15]. When the ΔS has a negative magnitude with low moisture contents, it is attributed to the existence of chemical adsorption and/or structural modifications of the adsorbent [29, 37].

3.2.4 *Net integral sorption enthalpy* (q_{eq}) ($J \cdot mol^{-1}$)

Changes in net integral enthalpy (Q_{in}) with the equilibrium moisture content are shown in **Figure 5**. This integral was determined the same way as the differential sorption enthalpy but at constant spreading pressure. The value of q_{eq} diminishes with increased X_w, varying on average for the three temperatures from 0.529 kJ · mol^{-1} for

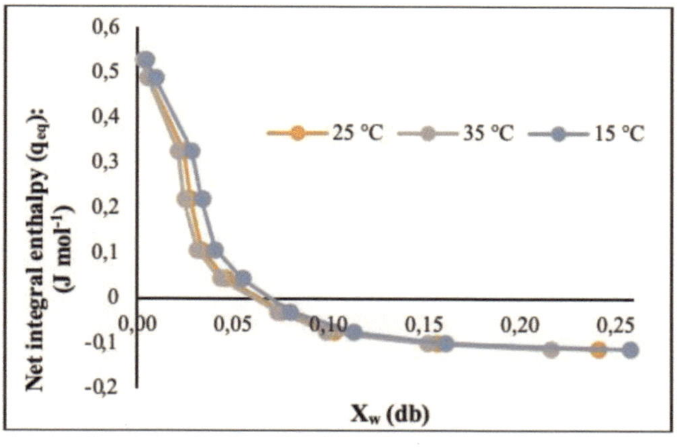

Figure 5.
Net integral enthalpy of the CP + FAC.

an X_w of 0.004 g of water/g s.s. at $-0.1106\ kJ \cdot mol^{-1}$ for 0.238 g of water/g s.s. This thermodynamic behavior may be attributed to the saturation of the active sorption points (moisture monolayer), given that water is adsorbed more easily in the active sites of the outer surface of the *CP + FAC* at low moisture contents, reaching the maximum value of q_{eq}. This maximum value indicates a greater water-*CP + FAC* interaction, which means that the sorption binding sites are stronger. Increased equilibrium moisture evidence diminished net integral enthalpy, which is attributed to the lesser favorability of the active sorption sites by being covered with water, forming multilayers. According to these results, variation in net integral enthalpy with X_w indicates that the level of solid-water interaction is higher than the interaction among water molecules, where the magnitude of enthalpy with high moisture content reflects the presence of free water. The determination of integral enthalpy is useful to quantify the energy required for drying, from the food's initial moisture content to its final moisture content. It also provides information on the microstructure of foods and the theoretical interpretation of physical phenomena that take place in the water-food interface.

A similar relation was found between q_{eq} and X_w in potato [41], concentrated freeze-dried yogurt [42], powdered pineapple pulp produced through different drying methods [43], potato starch [10], macaroni [29], chia seeds [4], and cassava flour [12].

3.2.5 Net integral sorption entropy (ΔS_{eq}) ($J \cdot mol^{-1} \cdot K^{-1}$)

Figure 6 shows the ΔS_{eq} in function of moisture content (X_w), finding that all values were negative, which could indicate existence of chemical adsorption or structural modification of the adsorbent or of the solid's outer surface, where ΔS_{eq} increased with the increase in equilibrium moisture content, varying from a minimum value of $-19.47\ J \cdot mol^{-1} \cdot K^{-1}$ for an X_w value of 0.005 g water/g s.s. at $-0.471\ J \cdot mol^{-1} \cdot K^{-1}$ for 0.257 g of water/g s.s. This result may be associated with greater mobility of the water molecules, favoring formation of multilayers. Hence, ΔS_{eq} tends to approach that of free liquid water. The minimum value of ΔS_{eq} may be considered the maximum stability point of foods during storage. Given that water molecules are strongly bound with the adsorbent in this zone, they are more neatly organized and less available to

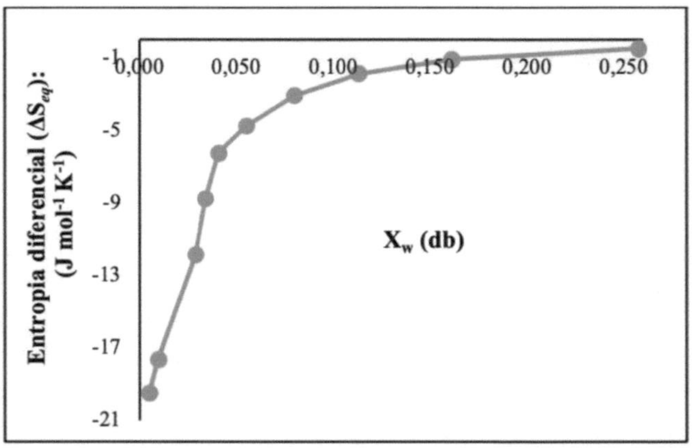

Figure 6.
Differential sorption entropy of the CP + FAC.

participate in deterioration reactions. Thereby, this minimum value of ΔS_{eq} is at an X_w close to the moisture of the monomolecular layer (m_o) predicted with the *BET* and *GAB* models (**Table 2**) [12, 13, 29].

3.3 Glass transition temperature (T_g) and its relation with the composition

Figure 7 shows the thermograms obtained through *DSC* for the *CP* + *FAC* stored at 35°C and *HR* = 65% in different a_w. In the range from 0.112 to 0.90, similar behavior was observed at 15 and 25°C (data not shown); additionally, T_g diminished when temperature was reduced. The temperature was taken from the midpoint of the change in heat capacity as the safest glass transition temperature (T_g), showing an inverse relation between T_g and a_w, that is, when increasing a_w, the T_g of the *CP* + *FAC* diminishes; this may be explained by the plasticizer effect of water because the addition of a plasticizer in the proximities of vitreous-amorphous powder reduces viscosity, increases molecular mobility, and increases the volume emptied by the "solid matter" provoking a volume available for its free movements (free volume) [1]. At the same time, it is noted that the thermograms obtained had a typical second-order transition, where the glass transition of amorphous materials produces a step change in the heat flow due to changes in heat capacity at phase transition temperature, which—as expected—the water plasticizer effect provokes significant decrease of T_g with increased X_w, where the water plasticizer effect may be based on weakening of hydrogen bonds and on dipole-dipole intra and inter macromolecular interactions due to shielding of these forces principally attractive by water molecules. This same behavior was found by other researchers who studied vitreous transitions and state diagrams for spray-dried tomato pulp [34], with mango powder [44], with tamarind seed mucilage [1], with freeze-dried pumpkin [6], and freeze dried pumpkin powders produced with different maltodextrin addition, where it was observed that T_{gs} increases with increasing maltodextrin content in the sample due to the addition of high molecular weight biopolymers, resulting in an improvement in its stability [8].

When increasing moisture content in the *CP* + *FAC*, it had great impact on fluidity, stickiness, surface adherence, and stability in storage because of its plasticizer effects and crystallization behavior.

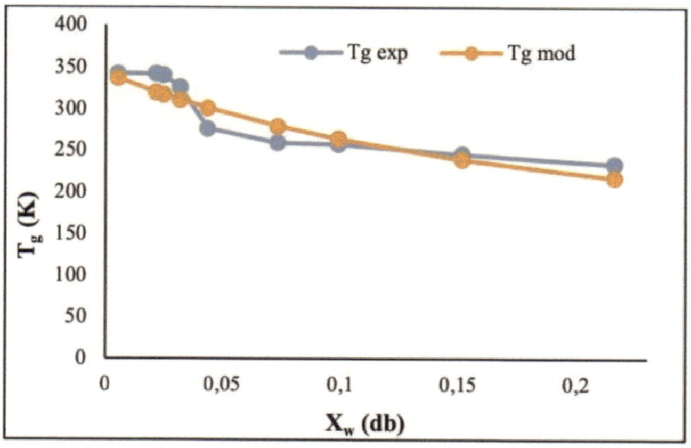

Figure 7.
DSC *analysis of the* CP + FAC *under different* a_w.

Availability of free water in a food component is expressed as a_w, responsible for the microbiological and biochemical reactions and, in turn, had great impact upon glass transition temperature (T_g), which varied for a_w: 0.112; T_g: 69.79°C, at values of a_w: 0.900; T_g: −39.0°C, demonstrating the plasticizer effect of water on the T_g, which was evident with the great reduction caused by increased moisture content (**Figure** 7) [4, 6, 33].

As the amount of plasticizer increases, the vitreous-amorphous material undergoes a change from the vitreous state to the gummy state. The effect of temperature on T_g is directly related with the capacity to adsorb moisture because when temperature increases, the attraction forces between molecules diminish due to increased kinetic energy of water molecules that leads to an increase in the distance among each other, and it is well-known that as moisture content is increased, T_g is decreased due to the low T_g of water (−137°C), which can be modeled by the Gordon-Taylor equation. Thus, the more effective measure of physical stability was the change in T_g as storage *RH* was increased, resulting in moisture content increase [1, 7, 8, 45].

The T_g value obtained for the *CP* + *FAC* with a_w: 0.112; with moisture content at 2.96%, it was 69.79°C, obtaining a high value because of the addition of maltodextrin as wall material whose T_g varies between 100 and 188°C, depending on its dextrose-equivalent (*DE*) property, said high T_g value (69.79°C) would not be achieved without using maltodextrin to obtain products with lower moisture and more stable [44].

The T_g of the *CP* + *FAC* was very close to that of products with starch, unlike other fruits, such as raspberry (T_g = 12.2°C) [38] and mango (T_g = 42.6°C) [46], which have more monosaccharides than polysaccharides with low molecular weight. Generally, T_g is related with the molecular weight (M_w). For example, the monosaccharide has a very low T_g, whereas the T_g of foods with high molecular weight that contain carbohydrates and proteins tends to be above 100°C [6].

Experimental data of T_g in the *CP* + *FAC*, balanced at nine levels of a_w, were fitted to the Gordon and Taylor model to predict the T_g of the binary solid mixture, where the predicted values and the fitted curve are shown in **Figure 8**, resulting in a reliable predictor model of the glass transition temperatures of the *CP* + *FAC*. The values of the parameters calculated through nonlinear regression were as follows: T_{gs} = 391.67 K and k = 0.753, with R^2 = 0.977, results somewhat different from those reported by

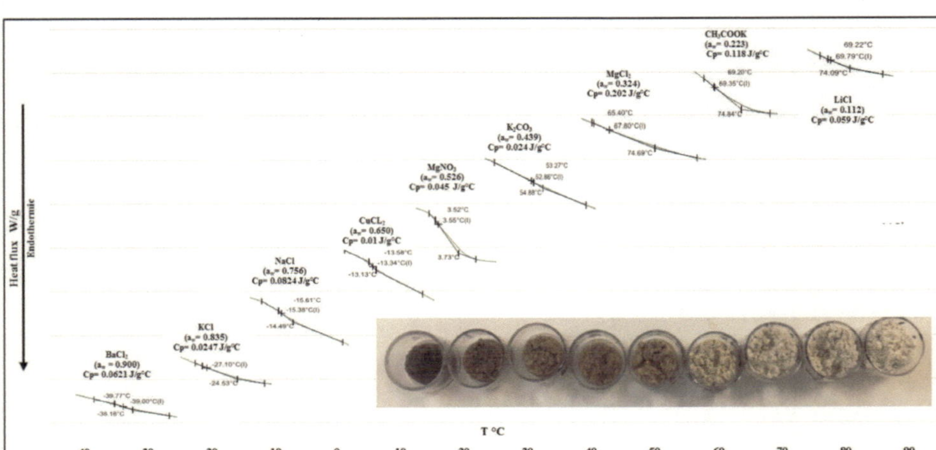

Figure 8.
Experimental glass transition (T_g) *curve and modeled through the Gordon and Taylor model in the* CP + FAC.

other authors, although these differences can be related with the different compositions of the food matrices, conditions of the drying process, and type of wall material used and concentration used [4, 6–8, 33, 34, 44], where the physical-chemical composition of the *CP* + *FAC* is fat = 30.54 ± 0.90%, protein = 4.07 ± 0.49%, total dietary fiber = 23.87 ± 1.61%, and 7% maltodextrin.

The *k* parameter controls the degree of curvature of the dependence of T_g on the water content (binary system) and may be related with the interaction force among the system's components. If $k = 1$, the glass transition temperature is a straight line. If $k > 1$, the nature of the graphic relation between the glass transition temperature and the content of solids is more concave; and for $k < 1$, the nature of the relation is convex [1, 4, 7, 8, 47].

3.4 Glass transition temperature (T_g) – water activity (a_w) relation and content in moisture (X_w – water activity (a_w) in the CP + FAC

Prediction of food stability based solely on data from sorption isotherms is not sufficient, given that certain physical-chemical and structural processes, such as stickiness, crunchiness, collapse, amorphous to crystalline transformations, and nonenzymatic browning rates are not related with the moisture value of the monolayer (m_o) and are best correlated with glass transition temperature (T_g) through plasticizing by water or temperature. Thus, using state diagrams that indicate the physical state of the material is combined with the sorption isotherms aids in the prediction of food stability regarding its physical characteristics. Several authors have coupled the data from sorption isotherms with those of T_g to obtain the critical conditions for food storage as follows: Tonon *et al.* [48], with acai powder; Djendoubi-Mrad *et al.* [47], with osmodehydrated apples and pears; Velázquez-Gutiérrez *et al.* [4], with chia seed mucilage; Zotarelli *et al.* [44], with mango powder. The critical content of water/water activity is the value at which the T_g of the product is equal to the room temperature. Above this temperature, vitreous and amorphous powders are susceptible to changes of spoilage, such as collapse, stickiness, and caking, which result in quality loss.

Thereby, to calculate the critical storage conditions of the *CP* + *FAC*, the sorption isotherms and data of T_g in function of a_w were plotted, and the critical values of a_w and X_w were obtained, considering a room temperature of 35°C (**Figure 9**); water content and T_g values were predicted by the *BET* - *GAB* and Gordon-Taylor models, respectively, observing that the *CP* + *FAC* had an a_{wc} of 0.46 and X_{wc} of 3.8% (*db*). This means that when the *CP* + *FAC* is stored at 35°C, the maximum relative humidity to which it can be exposed is 46%, and its moisture content would be 3.8%. When stored at higher relative humidity, the powder will undergo physical transformations, such as collapse, stickiness, and caking. These values are similar to those reported by Tonon *et al.* [48], with acai powder; but values much higher than those reported by Djendoubi-Mrad *et al.* [47], with osmodehydrated apples and pears, Zotarelli *et al.* [44], with mango powder, Al-Ghamdia *et al.* [6], with freeze-dried pumpkin, may be related with higher sugar content and acid present in these fruits compared with the *CP* + *FAC*.

3.5 Particle size distribution (D_{10}, D_{50}, and D_{90}) in the CP + FAC

The distribution of the average particle size in the three percentiles at time 0 (before storage) was as follows: D_{10}: 1.70 ± 0.05 μm; D_{50}: 8.46 ± 2.09 μm; D_{90}:

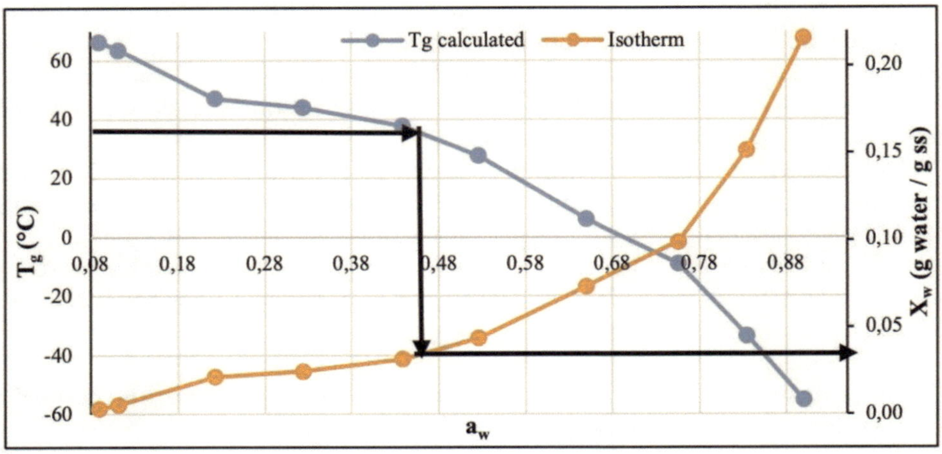

Figure 9.
Variation of glass transition temperature (T_g) *and the moisture sorption isotherm versus water activity* (a_w) *in the* CP + FAC *stored at 35°C.*

78.18 ± 24.30 μm (**Figure 10a**); after storage, agglomeration was generally present, forming larger particles. This may be attributed to the formation of bridges between the smaller particles, yielding an increase in average particle size in the three percentiles on day 180, thus, D_{10} = 2.04 ± 0.09 μm, D_{50} = 38.52 ± 3.67 μm, and D_{90} = 102.65 ± 6.82 μm (**Figure 10b**) [44].

This behavior could be explained by the continuous increase observed in moisture gain during storage, possibly causing significant agglomeration of the *CP + FAC*, losing its free-flowing characteristics and presenting clumping. This same behavior was reported by Pua *et al.* [49], with spray-encapsulated jackfruit oil (*Artocarpus heterophyllus*) and Mosquera-Mosquera [50], with borojó powder encapsulated with maltodextrin, indicating that strict control must be conducted of moisture content and storage at low temperatures, given that these are key factors to minimize agglomerate effects of powders. As a result of the humidification, sugars crystallize and form agglomerations because of the effect of stickiness. The collapse in pulverized products implies time-dependent structural changes, which are supposed to be changes in mechanical properties related with caking or stickiness. In addition, the rate of caking is in function of temperature, relative humidity, and storage time.

3.6 CP + FAC morphology

The microstructural analysis (*SEM*) of the *CP + FAC* particle morphology is a valid tool to determine and observe the agglomeration phenomenon or agglomeration on the powder surface during storage [51]. The microphotographs obtained at optimal drying conditions (day of storage 0) exhibited spherical shape and various sizes, which fluctuated between 20 and 65 μm, with smooth and rough surfaces, some with collapsed walls and agglomerate structures because of electrostatic effects and the van der Waals forces, characteristic of spray-dried powders, such as *CP + FAC* [44, 52–54] (**Figures 7** and **11a**). Spray drying generally shrinks the powder particles because of water vaporization, whereas freeze drying produces no shrinkage as water is frozen quickly in place followed by water sublimation, thus, retaining the same shape and

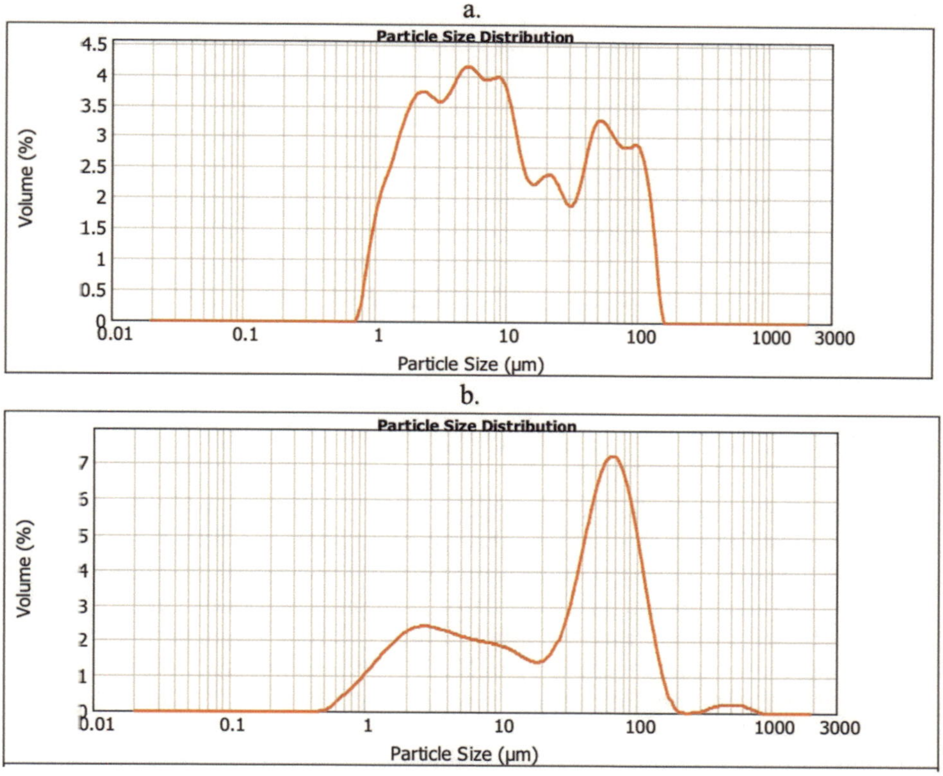

Figure 10.
Particle size distribution in the CP + FAC *stored at* $35°C$-N_2 *in times of storage (a) Day 0 and (b) day 180.*

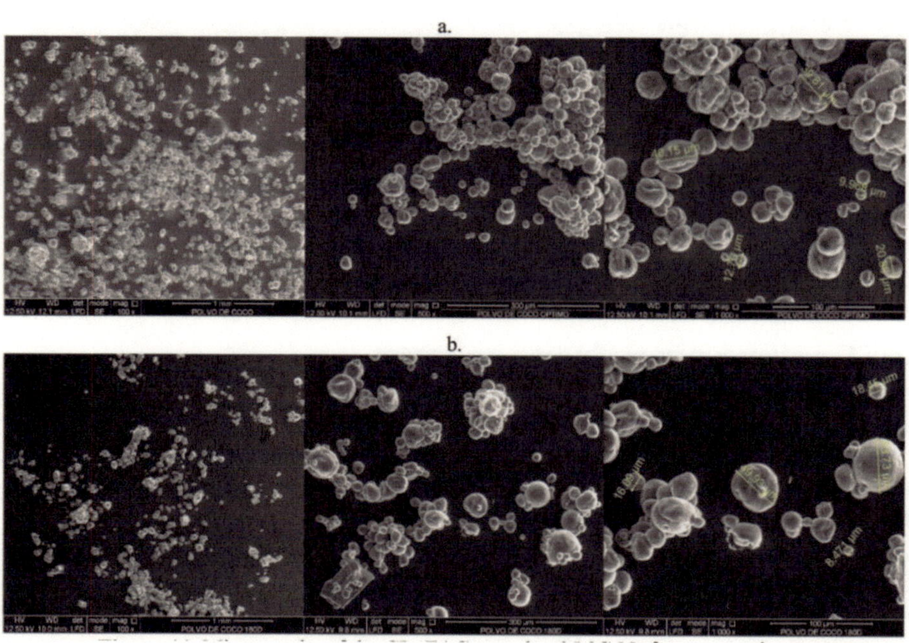

Figure 11.
Micrographs of the CP + FAC *stored at* $35°C$-N_2 *for storage times: (a) day 0 and (b) day 180.*

volume of particles [55]. Furthermore, the surface structure and the fibrous material without fractures or low-porosity matrix represent an indication of the effective role of maltodextrin as an encapsulating agent providing coverage over the nucleus that, likewise, acts as a thermal defense against oxidation and any unwanted physical and chemical change [44, 56, 57].

After 180 days of storage at 35°C-*N2* (**Figures** 7 and **11b**), it was found that the *CP* + *FAC* samples collapsed structurally because of the agglomeration phenomenon likely because of water adsorption on the particle surface, which could be the early stage of the caking phenomenon already described. Water absorption, together with T_g, plays a major role in the process of plasticization that causes the agglomeration of the sample. Water absorption leads to a decrease in the T_g of the sample [56–58].

4. Conclusions

This study determined the thermodynamic properties associated with moisture adsorption in the *CP* + *FAC*, where isosteric sorption heat calculated through the equation by Clausius-Clapeyron, increases by diminishing moisture content. Entropy changes were evident, showing that the adsorption process is irreversible. Spreading pressure increased with increased a_w and had little effect on temperature. It was observed that the net integral enthalpy of the *CP* + *FAC* diminishes with increasing moisture content. Net integral entropy increased continually with moisture content, but it had a negative value with respect to the entropy of pure liquid water. These adsorption findings in the *CP* + *FAC* are of interest in the determination of the optimal storage conditions, the prediction of shelf life, and the optimization and quantification of energy requirements during the transfer of mass and heat during drying.

The glass transition temperature (T_g) of the *CP* + *FAC*, stored in different water activities (a_w), diminished with increased water content, which varied for a_w: 0.112, T_g: 69.79°C at values of a_w: 0.900, T_g: −39.0°C, confirming the strong plasticizer effect of water upon this property. The critical activity (a_{wc}) and water content (X_{wc}) for the *CP* + *FAC* at 35°C were a_{wc} of 0.46 and X_{wc} of 3.8% (*db*), respectively; this means that when the *CP* + *FAC* is stored at 35°C, the maximum relative humidity to which it can be exposed is 46%, and its moisture content would be 3.8%. When stored at a higher relative humidity, the powder will undergo physical transformations, such as collapse, stickiness, and caking.

CRediT authorship contribution statement

Juan Carlos Lucas and Misael Cortes, Formal analysis. Conceptualization: Writing - review & editing. German Antonio Giraldo and Juan Carlos Lucas: Investigation, Writing - review & editing. Misael Cortes and Juan Carlos Lucas: Methodology, Resources, Writing - review & editing.

Funding

No funding was obtained for this study.

Conflict of interest

The authors declare that they have no known competing financial interests or personal relationships that could have appeared to influence the work reported in this paper.

Nomenclature

a_w	water activity
a_{wc}	critical water activity
A_m	area of water molecule: (1.06×10^{-19} m^2).
$a, b, c_1, c_2, m_1, m_2, n_1, n_2, A, B, C, K$	are fit parameters of each model
b and b_0	constants of Dent's sorption isotherm
db	dry base
X_w	moisture content
m_o	humidity of the monolayer
EMC	equilibrium moisture content
$CP + FAC$	coconut powder fortified with vitamins C, D_3, E and calcium
SD	spray dried
RH	relative humidity
R	constant of ideal gases ($KJ \cdot mol^{-1} \cdot K^{-1}$)
R^2	coefficient of determination
R_{MSE}	minimum root of mean square error
Q_{st}	Isosteric sorption heat (kJ mol^{-1})
ΔS	differential sorption entropy ($J \cdot mol^{-1} \cdot K^{-1}$)
Φ	spreading pressure ($J \cdot m^{-2}$)
q_{eq}	net integral sorption enthalpy ($J \cdot mol^{-1}$)
ΔS_{eq}	Net integral sorption entropy ($J \cdot mol^{-1} \cdot K^{-1}$)
T_g	glass transition temperature
X_{wc}	critical moisture content
DSC	differential scanning calorimetry
λ_{vap}	latent vaporization heat of pure water ($kJ \cdot mol^{-1}$)
T	absolute temperature (K)
X_{eq}	equilibrium moisture content ($kg \cdot kg^{-1}$ db)
K_B	Boltzmann's constant ($1.380 \times 10^{-23} J \cdot K^{-1}$)
T_{gw}	water glass transition temperature: −135.15°C (138.15 K)
T_{gs}	glass transition temperature of the anhydrous solids
x_w	mass fraction of water
k	constant of the Gordon-Taylor model
SEM	scanning electron microscopy
BET	Brunauer, Emmett and Teller mathematical model for modeling sorption isotherms
GAB	Guggenheim, Anderson and de Boer model for water sorption isotherm analysis

Author details

Juan Carlos Lucas-Aguirre[1*], German Antonio Giraldo-Giraldo[1] and
Misael Cortés-Rodríguez[2]

1 Facultad de Ciencias Agroindustriales, Universidad del Quindío, Armenia, Colombia

2 Departamento de Ingeniería Agrícola y Alimentos, Universidad Nacional de Colombia, Medellín, Colombia

*Address all correspondence to: jclucas@uniquindio.edu.co

References

[1] Alpizar-Reyes E, Carrillo-Navas H, Romero-Romero R, Varela-Guerrero V, Álvarez-Ramírez J, Pérez-Alonso C. Thermodynamic sorption properties and glass transition temperature of tamarind seed mucilage (*Tamarindus indica L.*). Food and Bioproducts Processing. 2017; **101**:166-176. DOI: 10.1016/j.fbp.2016.11.006

[2] Vasile FE, Judis MA, Mazzobre MF. Moisture sorption properties and glass transition temperature of a non-conventional exudate gum (*Prosopis alba*) from northeast argentine. Food Research International. 2020;**131**: 109033. DOI: 10.1016/j.foodres.2020.109033

[3] Carter BP, Schmidt SJ. Developments in glass transition determination in foods using moisture sorption isotherms. Food Chemistry. 2012;**132**:1693-1698. DOI: 10.1016/j.foodchem.2011.06.022

[4] Velázquez-Gutiérrez SK, Figueira AC, Rodríguez-Huezo ME, Román-Guerrero A, Carrillo-Navas H, Pérez-Alonso C. Sorption isotherms, thermodynamic properties and glass transition temperature of mucilage extracted from chia seeds (*Salvia hispanica L.*). Carbohydrate Polymers. 2015;**121**:411-419. DOI: 10.1016/j.carbpol.2014.11.068

[5] Mahato S, Zhu Z, Sun., D.W. Glass transitions as affected by food compositions and by conventional and novel freezing technologies: A review. Trends in Food Science & Technology. 2019;**94**:1-11. DOI: 10.1016/j.tifs.2019.09.010

[6] Al-Ghamdia S, Hong YK, Qu Z, Sablani SS. State diagram, water sorption isotherms and color stability of pumpkin (*Cucurbita pepo L.*). Journal of Food Engineering. 2020;**273**:109820. DOI: 10.1016/j.jfoodeng.2019.109820

[7] Vargas-Muñoz DP, Neves-deOliveira LA, Cardoso-daSilva L, Teixeira-Godoy H, Emy-Kurozawa L. Storage stability of 5-caffeoylquinic acid in powdered cocona pulp microencapsulated with hydrolyzed collagen and maltodextrin blend. Food Research International. 2020;**137**:109652. DOI: 10.1016/j.foodres.2020.109652

[8] Stępiénn A,Witczak M, Witczak T. Sorption properties, glass transition and state diagrams for pumpkin powders containing maltodextrins. LWT-Food Science and Technology. 2021;**134**: 110192. DOI: 10.1016/j.lwt.2020.110192

[9] Singh P, Talukdar P. Determination of thermophysical and desorption properties of elephant foot yam using composition based and fast sorption method. Thermal Science and Engineering Progress. 2020;**18**: 100508. DOI: 10.1016/j.tsep.2020.100508

[10] Al-Muhtaseb AH, McMinn WAM, Magee TRA. Water sorption isotherms of starch powders. Part 2: Thermodynamic characteristics. Journal of Food Engineering. 2004;**62**:135-142. DOI: 10.1016/S0260-8774(03)00202-4

[11] Ayala-Aponte A. Determinación de las isotermas de sorción y del calor isosterico en harina de yuca. Biotecnología en el Sector Agropecuario y Agroindustrial. 2011;**9**(1):88-96

[12] Ayala-Aponte A. Thermodynamic properties of moisture sorption in cassava flour. DYNA. 2016;**83**(197):

139-145. DOI: 10.15446/dyna.v83n197.51543

[13] Caballero-Gutiérrez BL, Márquez-Cardozo CJ, Ciro-Velásquez HJ. Thermodynamic study of adsorption properties of Rocoto Pepper (*Capsicum pubescens*) obtained by freeze-drying. Advance Journal of Food Science and Technology. 2018;**15**(SPL):91-98. DOI: 10.19026/ajfst.15.5877

[14] Moussaoui H, Bahammoua Y, Idlimama A, Abdenouri N. Investigation of hygroscopic equilibrium and modeling sorption isotherms of the argan products: A comparative study of leaves, pulps, and fruits. Food and Bioproducts Processing. 2019;**114**:12-22. DOI: 10.1016/j.fbp.2018.11.002

[15] Largo-Ávila E, Cortes-Rodríguez M, Ciro-Velásquez HJ. The adsorption thermodynamics of sugarcane (*Saccharum officinarum L.*) powder obtained by spray dryig technology. VITAE. 2014;**21**(3):165-177

[16] Lucas-Aguirre JC, Tobón-Castrillón C, Cortes-Rodríguez M. Influence of the composition of coconut-based emulsions on the stability of the colloidal system. Advance Journal of Food Science and Technology. 2018a;**14**(3):77-92. DOI: 10.19026/ajfst.14.5841

[17] Lucas-Aguirre JC, Giraldo-Giraldo GA, Cortes-Rodríguez M. Effect of the spray drying process on the quality of coconut powder fortified with calcium and vitamins C, D_3 and E. Advance. Journal of Food Science and Technology. 2018b;**16**(SPL):102-124. DOI: 10.19026/ajfst.16.5943

[18] Lucas-Aguirre JC, Giraldo-Giraldo GA, Cortes-Rodríguez M. Modelling of moisture sorption isotherm and evaluation of net isosteric heat for spray-dried fortified coconut (*Cocos nucifera L.*) powder. British Food Journal. 2021; **127**(7):2349-2364. DOI: 10.1108/BFJ-11-2020-1017

[19] Van den Berg C, Bruin S. Water activity and its estimation in food systems. In: Rockland LB, Stewart GF, editors. Water Activity: Influences on Food Quality. New York: Academic Press; 1981. pp. 1-61

[20] Brunauer S, Emmett PH, Teller E. Adsorption of gases in multi-molecular layers. Journal of American Chemistry Society. 1938;**60**:309-320. DOI: 10.1021/ja01269a023

[21] Oswin CR. The kinetics of package life III. The isotherm. Journal of Chemical Industry. 1946;**65**:419-421. DOI: 10.1002/jctb.5000651216

[22] Smith SE. The sorption of water vapor by high polymers. Journal of the American Chemical Society. 1947;**69**: 646. DOI: 10.1021/ja01195a053

[23] Halsey G. Physical adsorption on non-uniform surfaces. The Journal of Chemical Physics. 1948;**16**:931-937

[24] Henderson S. A basic concept of equilibrium moisture. Agricultural Engineering. 1952;**33**:29-32

[25] Chung DS, Pfost H. Adsorption and desorption of water by cereal grains and their products. Transactions of the American Society of Agricultural Engineering. 1967;**10**:52-555. DOI: 10.13031/2013.39726

[26] Peleg M. Assessment of a semi-empirical four parameter general model for sigmoid moisture sorption isotherms. Journal of Food Process Engineering. 1993;**16**:21-37. DOI: 10.1111/j.1745-4530.1993.tb00160.x

[27] Caurie M. A new model equation for predicting safe storage moisture levels for optimum stability of dehydrated foods. Journal of Food Technology. 1970; **5**:301-307. DOI: 10.1111/j.1365-2621.1970.tb01571.x

[28] Kaya S, Kahyaoglu T. Thermodynamic properties and sorption equilibrium of pestil (grape leather). Journal of Food Engineering. 2005;**71**: 200-207. DOI: 10.1016/j.jfoodeng.2004.10.034

[29] Arslan N, Toğrul H. Modelling of water sorption isotherms of macaroni stored in a chamber under controlled humidity and thermodynamic approach. Journal of Food Engineering. 2005;**69**: 133-145. DOI: 10.1016/j.jfoodeng.2004.08.004

[30] Kamau E, Mutungi C, Kinyuru J, Imathiu S, Tanga C, Affognon H, et al. Moisture adsorption properties and shelf-life estimation of dried and pulverised edible house cricket *Acheta domesticus (L.)* and black soldier fly larvae *Hermetia illucens (L.)*. Food Research International. 2018;**106**: 420-427. DOI: 10.1016/j.foodres.2018.01.012

[31] Dent RW. A multilayer theory for gas sorption. Part 1: Sorption of a single gas. Textile Research Journal. 1977;**47**: 145-152

[32] Grabowski JA, Truong VD, Daubert DR. Spray drying of amylase hydrolyzed sweetpotato puree and physicochemical properties of powder. Journal of Food Science. 2006;**71**:E209-E217. DOI: 10.1111/j.1750-3841.2006.00036.x

[33] Islam MZ, Kitamura Y, Yamano Y, Kitamura M. Effect of vacuum spray drying on the physicochemical properties, water sorption and glass transition phenomenon of orange juice powder. Journal of Food Engineering. 2016;**139**:131-140. DOI: 10.1016/j.jfoodeng.2015.08.024

[34] Goula AM, Karapantsios TD, Achilias DS, Adamopoulos KG. Water sorption isotherms and glass transition temperature of spray dried tomato pulp. Journal of Food Engineering. 2008;**85**: 73-83. DOI: 10.1016/j.jfoodeng.2007.07.015

[35] Mirhosseini H, Ping Tan C, Hamid NSA, Yusof S. Effect of Arabic gum, xanthan gum and orange oil contents on ζ-potential, conductivity, stability, size index and pH of orange beverage emulsion. Colloids and Surfaces A: Physicochemical Engineering Aspects. 2008;**315**:47-56. DOI: 10.1016/j.colsurfa.2007.07.007

[36] Cano-Chauca M, Stringheta PC, Ramos AM, Cal-Vidal J. Effect of the carriers on the microstructure of mango powder obtained by spray drying and its functional characterization. Innovative Food Science Emerging. 2005;**6**:420-428. DOI: 10.1016/j.ifset.2005.05.003

[37] Iglesias HA, Chirife J, Ferro-Fontan C. On the temperature dependence of isosteric heats of water sorption in dehydrated foods. Journal of Food Science. 1989;**54**(6):1620-1631

[38] Syamaladevi RM, Sablani SS, Tang J, Powers J, Swanson BG. State diagram and water adsorption isotherm of raspberry (*Rubus idaeus*). Journal of Food Engineering. 2009;**91**:460-467. DOI: 10.1016/j.jfoodeng.2008.09.025

[39] Duarte-Marques RC, Resende-Oliveira É, Mendes-Coutinho GS, Chaves-Ribeiro AE, Souza-Teixeira C, Soares-Júnior MS, et al. Modeling sorption properties of maize by-products obtained using the Dynamic Dewpoint

Isotherm (DDI) method. Food Bioscience. 2020;**38**:100738. DOI: 10.1016/j.fbio.2020.100738

[40] Ajibola OO, Aviara NA, Ajetumobi OE. Sorption equilibrium and thermodynamic properties of cowpea (*Vigna unguiculata*). Journal of Food Engineering. 2003;**58**:317-324. DOI: 10.1016/S0260-8774(02)00394-1

[41] McMinn W, Magee T. Thermodynamic properties of moisture sorption of potato. Journal of Food Engineering. 2003;**60**(2):157-165. DOI: 10.1016/S0260-8774(03)00036-0

[42] Azuara E, Beristain CI. Enthalpic and entropic mechanisms related to water sorption of yogurt. Drying Technology. 2006;**24**:1501-1507. DOI: 10.1080/07373930600961173

[43] Viganó J, Azuara E, Telis VRN, Beristain CI, Jiménez M, Telis-Romero J. Role of enthalpy and entropy in moisture sorption behavior of pineapple pulp powder produced by different drying methods. Thermochimica Acta. 2012; **528**:63-71. DOI: 10.1016/j.tca.2011.11.011

[44] Zotarelli MF, Martins-da Silva V, Durigon A, Dupas-Hubinger M, Borges-Laurindo J. Production of mango powder by spray drying and cast-tape drying. Powder Technology. 2017;**305**:447-454. DOI: 10.1016/j.powtec.2016.10.027

[45] Voelker AL, Sommer AA, Mauer LJ. Moisture sorption behaviors, water activity-temperature relationships, and physical stability traits of spices, herbs, and seasoning blends containing crystalline and amorphous ingredients. Food Research International. 2020;**136**: 109608. DOI: 10.1016/j. foodres.2020.109608

[46] Zhao JH, Liu F, Wen X, Xiao HW, Ni YY. State diagram for freeze–dried mango: Freezing curve, glass transition line and maximal–freeze–concentration condition. Journal of Food Engineering. 2015;**157**:49-56. DOI: 10.1016/j. jfoodeng.2015.02.016

[47] Djendoubi-Mrad N, Bonazzi C, Courtois F, Kechaou N, Boudhrioua-Mihoubid N. Moisture desorption isotherms and glass transition temperatures of osmo-dehydrated apple and pear. Food and Bioproducts Processing. 2103;**91**:121-128. DOI: 10.1016/j.fbp.2012.09.006

[48] Tonon RV, Baroni AF, Brabet C, Gibert O, Pallet D, Hubinger MD. Water sorption and glass transition temperature of spray dried açai (*Euterpe oleracea Mart.*) juice. Journal of Food Engineering. 2009;**94**:215-221. DOI: 10.1016/j.jfoodeng.2009.03.009

[49] Pua CK, Hamid NSA, Tan CP, Mirhosseini H, Rahman RA, Rusul G. Storage stability of jackfruit (*Artocarpus heterophyllus*) powder packaged in aluminium laminated polyethylene and metallized co-extruded biaxially oriented polypropylene during storage. Journal of Food Engineering. 2008; **89**(4):419-428. DOI: 10.1016/j. jfoodeng2008.05.023

[50] Mosquera-Mosquera LH. Influencia de la humedad y de la adición de solutos (maltodextrina o goma arábiga) en las propiedades fisicoquímicas de borojó y fresa en polvo. Tesis Doctoral. Repositorio institutional UPV (Universidad Politécnica de Valencia); 2010. p. 219

[51] Liu F, Cao X, Wang H, Lia X. Changes of tomato powder qualities during storage. Powder Technology. 2010;**204**:159-166. DOI: 10.1016/j. powtec.2010.08.002

[52] Frascareli EC, Silva VM, Tonon RV, Hubinger MD. Effect of process

conditions on the microencapsulation of coffee oil by spray drying. Food and Bioproducts Processing. 2012;**90**: 413-424. DOI: 10.1016/j.fbp.2011.12.002

[53] Jafari SM, Ghalenoei MG, Dehnad D. Influence of spray drying on water solubility index, apparent density, and anthocyanin content of pomegranate juice powder. Powder Technology. 2016; **311**:59-65. DOI: 10.1016/j.powtec.2017.01.070

[54] Mohammed NK, Tan CP, Manap YA, Alhelli AM, Hussin ASM. Process conditions of spray drying microencapsulation of *Nigella sativa* oil. Powder Technology. 2017;**315**:1-14. DOI: 10.1016/j.powtec.2017.03.045

[55] Deshwal GK, Kumar-Singh A, Kumar D, Sharma H. Effect of spray and freeze drying on physico-chemical, functional, moisture sorption and morphological characteristics of camel milk powder. LWT – Food Science and Technology. 2020;**134**:110117. DOI: 10.1016/j.lwt.2020.110117

[56] Zhang L, Qiu J, Cao X, Zeng X, Tang X, Sun Y, et al. Drying methods, carrier materials, and length of storage affect the quality of xylooligosaccharides. Food Hydrocolloids. 2019;**94**:439-450. DOI: 10.1016/j.foodhyd.2019.03.043

[57] Mutlu C, Koç A, Erbaş M. Some physical properties and adsorption isotherms of vacuum-dried honey powder with different carrier materials. LWT - Food Science and Technology. 2020;**134**:110166. DOI: 10.1016/j.lwt.2020.110166

[58] Islam-Shishir MR, Taip FS, Saifullah M, Aziz N, Talib RA. Effect of packaging materials and storage temperature on the retention of physicochemical properties of vacuum-packed pink guava powder. Food Packaging and Shelf Life. 2017;**12**:83-90. DOI: 10.1016/j.fpsl.2017.04.003

Section 2

Novel High-Efficiency Sorbents

Chapter 5

Bottle Gourd (*Lagenaria vulgaris*) Shell as a Natural, Biodegradable, Highly Available, Cheap, Agricultural by-Product, Miscellaneous Biomass, Ion Exchanger, Biosorbent and Fertilizer

Goran Nikolić, Dragana Marković Nikolić, Aleksandar Bojić, Danijela Bojić, Ljubiša Nikolić, Ljiljana Stanojević, Miloš Durmišević, Nataša Simonović and Miloš Kostić

Abstract

The increased interest in natural, renewable, biodegradable, easily available, low-cost materials makes agricultural residues, such as lignocellulosic biomass, attractive raw materials for the preparation of effective biosorbents for various pollutants (metal ions, anions, dyes, pharmaceutical degradation products, metabolites, organic macromolecules) for the wastewater treatment. Various covalent and non-covalent modification approaches significantly improve the sorption properties of these lignocellulosic functional particles, even improving their dispersion in hydrophobic polymer matrices, associative properties in water, rheological properties, surface-active properties, which can control the sorption of various ionic pollutants both in batch and in flow mode. Advantages over commercial sorbents (techno-economic aspect, no secondary pollution, usability as fertilizers), easy separation from the sorption medium, microstructural properties (strength, porosity, interactivity, stability), as a promising and sustainable biosorbent highlight the environmentally friendly bottle gourd shell. On the example of this biosorbent, the conventional approach to the pollutant sorption process (comparative kinetic, thermodynamic and equilibrium tests) was improved, as well as its shortcomings in predicting optimal process parameters. To fill the gaps of the already unnecessary numerous experiments, a design study involving OVAT experimental approaches integrated with DoE methodology was conducted. This integrated experimental design was implemented in the optimization of the pollutant sorption process.

Keywords: *Lagenaria vulgaris*, bottle gourd shell, biosorbent, ion exchanger, pollutants

1. Introduction

The generation of wastewater on a large scale is an inevitable consequence of today's modern society. One of the reasons for this environmental problem is the rapid growth of the human population, which has a significant impact on the natural waters pollution by municipal waste. On the other hand, the progressive development of industry has led to increased production of industrial wastewater and its discharge into natural waterways. The generation of municipal and industrial wastewater, polluted with organic and inorganic pollutants, is becoming an increasing problem for human health and the environment.

Various conventional methods are used for wastewater treatment (chemical precipitation, oxidation-reduction, coagulation and flocculation, sedimentation). More efficient, but also more expensive procedures have been developed, such as: ion exchange, reverse osmosis, electrodialysis, ultrafiltration, membrane separation and adsorption with activated carbon. Due to its advantages, such as ease of use and the efficiency of removing various pollutants, adsorption with activated carbon has become one of the most commonly applied wastewater treatments. However, due to the high costs of obtaining and the complexity of the regeneration process, which causes secondary pollution, the application of activated carbon is limited today.

In order to overcome the shortcomings of this adsorbent, numerous studies have shown that various agro-industrial waste biomasses can be used alternatively for these purposes [1, 2]. There are well-known examples of sorbents based on: tree bark, fruit peels (banana, orang, lemon, grapefruit), woody shell (walnut, peanut, almond), pits (plum, cherry, peach), seed husks (rice, wheat, pumpkin, sunflower), straw, algae and others. Of particular importance is the pretreatment of biomass, for which numerous technologies based on physical, chemical and biological methods have been developed [3].

Apart from native biomass as cheap natural lignocellulosic polymers, sorbents that can be obtained through a series of simple physical and chemical modifications of waste biomass are of significant interest for these purposes [4–6]. Today, biomass modification is increasingly the subject of scientific research due to improved characteristics and their potential application in numerous fields. Modified sorbents based on lignocellulosic material have an advantage compared to synthetic polymers, considering that they are significantly cheaper, renewable and less toxic [7]. In addition, they represent a functional ecological material due to unique properties, such as: hydrophilicity, adsorptivity, biocompatibility and biodegradability.

Modification of lignocellulosic material, by introducing specific functional groups, is of interest for obtaining sorbents with improved characteristics. For example, cationic functional groups (from quaternary ammonium salts) increase the sorption capacity for anions, primarily the ion exchange capacity [8]. Sorbents with cationic active centers are of particular importance for the treatment of wastewater containing large amounts of various anionic pollutants (nitrates, phosphates, sulfates, cyanides, etc.).

It is characteristic that municipal and agro-industrial waste waters increasingly contain excessive concentrations of phosphates and nitrates. These anionic species represent a serious problem of endangering the environment (they are directly responsible for the occurrence of eutrophication), and can also lead to serious health problems. Considering that these anionic species are present in water in the form of soluble salts, conventional procedures for water treatment (coagulation, sedimentation) have not proven to be effective for their removal. Developed physical-chemical (reverse osmosis, electro-dialysis) and biological denitrification methods [9], as well as the application of commercial ion-exchange resins for the removal of nitrates and phosphates from wastewater [10], are not sustainable from the economy aspect. Therefore, further research was focused on sorption processes, using cheap agricultural waste as ecological sorption material. Various studies have investigated the cationic modification of biomass via quaternization of lignocellulosic material with amino [1, 2] or ammonium compounds [11], as well as surfactants [12]. In this sense, various agricultural by-products were examined: wheat straw, sugar cane, rice husk, corn grain, sawdust, etc. [13–16].

Bearing in mind the mentioned facts and insufficiently solved current ecological problems, the development of new cationic sorbents will always be an open topic of research. First of all, these researches are based on the discovery of new usable sorbents of natural origin, as well as the improvement of their sorption characteristics. In this sense, the attention of our research group during the last decade was attracted by the fact that in the literature there was no data on the use of the *Lagenaria vulgaris* shell (LVS) as a sorbent of anionic pollutants. Therefore, as a contribution to this field, our research was focused on the development of new cationic sorbents through the lignocellulosic biomass modification of LVS, a plant by-product that had not been used for these purposes until then.

Namely, due to its nutritional and medicinal properties, the bottle gourd (*Lagenaria vulgaris* species) is increasingly cultivated in many countries of the world. It is a good prerequisite for the availability of a significant amount of LVS as a potentially valuable agro-waste [17–19]. The bottle gourd shell itself is characterized by a lignocellulosic composition, an extremely compact and solid structure in the mechanical sense, as well as a porous structure, with low ash content and a high volumetric mass. Tests have shown that LVS biomass, as well as its modified products, do not swell in water and are easily separated from the water phase after treatment [20]. These are very important properties of a precursor for the sorbents synthesis, which are not found in most other lignocellulosic biomasses. Additionally, LVS lignocellulosic biomass contains hydroxyl groups on the glucopyranose units of cellulose as the main functional groups, which are easily susceptible to various chemical reactions. In order to synthesize cationic sorbents, research into the chemical modification of LVS biomass was focused on the ammonolysis of alkaline treated or etherified cellulose, as well as the grafting of a cationic surfactant onto the lignocellulosic skeleton of biomass [21]. The obtained cationically modified products were expected, first of all, to have a high sorption capacity for various anionic species. In this sense, the optimization of the synthesis process parameters and the definition of the product sorption capacity towards nitrates and phosphates from the aqueous solution are of particular importance. Feedback on the influence of process parameters on the formation, amount and type of active sorption centers is important for improving the efficiency and selectivity of the anionic species sorption process. Sorption tests of nitrates and phosphates in a mixture with other anionic species can be helpful in the application of these cationic sorbents in real conditions of municipal or industrial wastewater treatment.

From all of the above, potential sorbents based on LVS have become the main subject of our research. Initially, LVS in its native form [20] or in some chemically modified forms (xanthated, sulfonated, and methylsulfonated) [22–25] was investigated as a sorbent for the removal of heavy metals from water. Later, research was focused on the removal of other pollutants from aqueous solutions, such as dyes [26], herbicides and pharmaceutical substances [27]. More recent research is related to the modification of sorbents to solve the problem of eutrophication, primarily the removal of phosphates and nitrates, which is the subject of this book chapter.

Considering the development and improvement of new sorbent materials, in addition to the expected scientific contribution, the research presented in this chapter has great practical importance from both the technological and ecological aspects. Due to changes in the structure and chemical composition of the starting raw material, as well as the predicted ion exchange mechanism, the advantage of the obtained sorbents is the possibility of multiple regeneration and reuse. On the other hand, it solves the problem of disposal of anion-saturated biomass, which can be composted and used for fertilizing agricultural land with renewable nutrients (phosphates and nitrates).

2. Bottle gourd (*Lagenaria vulgaris*)

The bottle gourd (or flask gourd) is one of the 6 known species of the gourd genus *Lagenaria*. The calabash or *Lagenaria siceraria* (synonym *Lagenaria vulgaris* Ser.) is the best-known species of aerial annual creeping plant from the kingdom *Plantae* (division *Magnoliophyta*), squash family *Cucurbitaceae*, genus *Lagenaria* (*Cucurbita lagenaria* L.) [28]. This species is one of the first cultivated plants in the world (indigenous species), mainly for its fruit. It grows on sandy soils and red loam, on flat surfaces and moderate slopes of the ground [17]. Given that *Lagenaria vulgaris* requires a lot of light and heat, it originates from Africa or Asia (morphologically different autochthonous species), from where it was transferred to other countries of the world. Apart from bottle gourd, other species are not cultivated.

2.1 Bottle gourd valorization

Young fruits of *Lagenaria vulgaris* (LV) have light-green smooth skin and spongy white pulp, in which the seeds are arranged (**Figure 1**). Harvested young gourds are used as vegetables. Mature gourd in the shape of a bottle reaches a size of about 50 × 20 cm, and a weight of up to 1.2 kg. The gourd shell is hard and woody, and very bitter in taste. After matures, the gourd dries completely naturally, so only the seed remains in the fruit. As the shell matures and dries, it acquires a yellow or light-brown color.

The *Lagenaria vulgaris* is a plant of economic importance, considering that the pulp has nutritious and pharmaceutical properties [18]. The pulp contains 85–95% water and is low in calories (20 kcal/100 g). In addition to water, it also contains some

Figure 1.
Lagenaria vulgaris: Flower (a), young fruit (b), spongy tissue with seeds (c), dried pulp (d) and dried gourd shell (e).

carbohydrates, proteins, vitamins A, B and C, as well as minerals (calcium and iron). In traditional medicine, the pulp is known as a medicine for the treatment of skin diseases. The pulp can be used as a sedative and emetic purgative, diuretic, as well as for pectoral complaints. In addition, phytochemical and pharmacological studies have indicated the potential treatment of heart failure, hypertension, bilious disorders (jaundice), diabetes, ulcers, hemorrhoids, etc. [18]. Extracts of this plant have been proven to have antioxidant and antibiotic activity [19].

Lagenaria vulgaris is cultivated not only for its nutritional and medicinal properties but also for the practical use of the dried shell (LVS). After ripening and drying, the shell is used for various purposes, such as water storage vessels (the so-called "lejka" in Serbian), funnel, pipes, accessories (musical instruments, liquid storage containers).

2.2 Physico-chemical properties of bottle gourd shell

The bottle gourd shell does not retain a large amount of water in its structure (moisture content 3.8%), which is why the crushed material can be stored in the air without sticking particles or changing granulation. The mineral substances of the shell are mainly alkaline earth metals (Ca and Mg), which the plant accumulates during its growth. This inorganic fraction is mostly incorporated into the highly developed porous structure of biomass and is found in conjunction with surface functional groups. The gourd shell is characterized by relatively low values of density and bulk weight, which is important for application as a sorbent [20].

Based on Boehm's method for determining the proportion of functional groups, the concentration of strongly acidic groups of 0.138 mmol/g and very weakly acidic groups of 0.104 mmol/g was determined in the LVS native form. The amount of weakly acidic groups is slightly higher and is 0.218 mmol/g [26]. The specific surface area and porosity of the LVS were analyzed using the isotherm of the nitrogen sorption/desorption process [29]. The analysis showed that nitrogen saturation is achieved at relatively low gas pressure values. The small value of the sorbed gas volume at the moment of equilibrium (0.28 cm^3/g) indicated a small specific surface area of the material, without pronounced porosity. Using the BET method, a total specific surface of 1.075 m^2/g was determined, whereby the outer surface of the shell (0.226 m^2/g) has a relatively small share in the total surface. The shell surface is characterized mainly by micropores (1.048 m^2/g) of small volume ($4.96 \cdot 10^{-4}$ cm^3/g), while the remaining specific surface is represented by macropores and surface elements of significantly larger dimensions than standard pores (cavities and channels). The presence of macropores and larger cavities can enable better movement of the aqueous phase through the structure of the material and promote internal diffusion [30].

The constitutional analysis of the LVS structural components showed that this lignocellulosic biomass mainly consists of holocellulose (cellulose and hemicellulose) and lignin [21, 29]. These are typical structural constituents of woody plants or the bark of many other similar plant products (pumpkin, corn on the cob, chestnut, hazelnut, peanut and others) [1]. Cellulose and hemicellulose are generally classified as polysaccharides, while lignin is a complex phenolic (aromatic) polymer. Cellulose does not dissolve in water, diluted acids and bases. It has a well-ordered (crystalline) and randomly ordered (amorphous) structure. This property is of particular importance, because it makes the cellulose material chemically and mechanically stable and resistant to aqueous solutions. Hemicellulose has a more complex structure (pentose-hexose composition) and bonds than cellulose. It is linked to the cellulose microfibrils by non-covalent bonds in an amorphous matrix, thus holding the rigid cellulose fibers in place. As a polysaccharide of amorphous nature (due to branching), hemicellulose is sensitive to acids and bases in aqueous solution and more susceptible to depolymerization. Lignin is a macromolecule built from phenylpropane units derived from p-coumaryl, coniferyl and sinapyl alcohol as monomers. The composition of macromolecules includes methoxyl, acetyl and formyl groups. As an amorphous polyaromatic substance, in addition to the basic phenylpropanoid units, lignin also contains aliphatic constituents that affect the increase in hydrophobicity. Of the 11 types of intermolecular bonds within the molecule, the most abundant is the β-O-4 ether bond (60%), which leads to linear stretching of the polymer. Lignin is stable in contact with water, weak bases and acids. However, it undergoes chemical changes under the influence of elevated temperature. The three-dimensional structure of lignin, interwoven with cellulose and hemicellulose molecules, gives LVS biomass distinct hardness and mechanical stability. Due to the strong mutual interaction of phenylpropanoid units with other lignin molecules and polysaccharides, difficulties arise in the isolation of lignin from biomass. Therefore, for applying any method of delignification or changing its structure, the inevitable breaking of strong covalent bonds must be taken into account.

Such structural properties of lignocellulosic biomass are of crucial importance for the direct application of LVS as a sorbent of various pollutants, primarily heavy metals or other cationic species from waste or contaminated natural waters [27]. Considering the structural characteristics, LVS in contact with water does not swell and does not change its macroscopic or microscopic form. Also, after separation from the aqueous phase, LVS is very easily washed and dried, which is important for its regeneration and reuse. Both structural components (cellulose and lignin) contain a large number of different oxygen functional groups, which are responsible for binding cationic pollutants [20].

In addition, during the physico-chemical treatment, the LVS retains the solid nature of the structure, built of cellulose and lignin, which is characterized by mechanical resistance and compactness as important characteristics for the LVS application as a sorbent.

2.3 Sorption characteristics of bottle gourd shell

Raw LVS biomass does not possess sorption properties or exhibits them to a very small extent (<1% of various sorbates), which is not of practical importance for these purposes [21]. However, modified LVS biomass can be a good sorbent for a wider range of pollutants. Like other lignocellulosic biomasses, LVS can be modified by chemical treatment or thermal carbonization. Biomass modified with various chemical agents is suitable for removing cationic pollutants from water, such as heavy metals and cationic dyes. Biomass modified by cationic grafting of cellulose chains is important for the purification of waters contaminated with anionic species. Activated carbon obtained by thermal carbonization of biomass and steam activation is a good sorbent of organic, non-polar and weakly polar pollutants, such as some pharmaceutical substances and pesticides.

Based on numerous methods of chemical modification, such as acid-base activation [20], sulfonation [22], methyl-sulfonation [23], xanthation [24], several procedures have been developed to obtain new, more efficient and economical LVS sorbents. The obtained materials were characterized in detail by various physico-chemical methods and applied as sorbents for the removal of various cationic species from aqueous solutions in a batch mode. The results of our research confirmed that sorbents based on modified bottle gourd shell can be used for effective removal of heavy metals, primarily Pb(II), Cd(II), Zn(II), Cu(II), Ni(II) [20, 25, 31]. Also, we confirmed the sorption efficiency of other cationic pollutants from natural or wastewaters, such as the cationic dye methylene blue [26]. These sorbents represent an alternative to more expensive commercial sorbents based on activated carbons, as well as other sophisticated technologies.

The possibility of using LVS biomass as a sorbent for some drugs was investigated [30, 32], given their significant presence in municipal waters via the sewage system. Namely, drugs and their metabolites are excreted from the body through urine and feces, so it is necessary to remove them from wastewater before discharge into natural waterways (rivers and lakes). One such drug is ranitidine hydrochloride (RH), which is classified as a drug with hazardous effects on aquatic organisms. Additionally, RH is partially biodegradable and susceptible to structural change under the influence of sunlight, creating toxic photoproducts. In this sense, the sorption removal of ranitidine hydrochloride from an aqueous solution was monitored using *Lagenaria* activated carbon (LAC). Tests were performed in batch mode, in the presence and absence of ultrasound. LAC is characterized as a porous spongy material with an amorphous structure, high surface area (665 m^2/g) and pH_{PZC} (point of zero charge) of 7.2. The process of RH removal is best described by the pseudo-second order kinetic model and the Langmuir isotherm, which indicated monolayer sorption and significantly faster equilibrium establishment in the presence of ultrasound. The maximum RH sorption capacity of 425 mg/g for ultrasound power of 50 W was achieved at ambient temperature. Thermodynamic data indicated that RH sorption is an exothermic and spontaneous process, implying a physical mechanism of RH removal from aqueous solutions by LAC. The obtained data were used to design a sorption process in order to remove other organic and weakly polar pollutants from aqueous solutions, such as herbicide 2,4-dichlorophenoxyacetic acid [27].

From an economical aspect, LVS as a precursor for the production of a superior carbon sorbent can be considered an extremely economical resource because the costs of its production are minimal [30]. Namely, the LV plant is grown without special requirements, without the use of pesticides and expensive agricultural procedures. On the other hand, the thermochemical conversion to the final carbon product consists only of the costs of a few cheap chemicals and the electricity used for heating. The final cost of the obtained LVS carbon is estimated at less than 500 USD/t, in contrast to commercial activated carbons present on the market in the form of powders and granules whose average value ranges between 600 and 2000 USD/t. Therefore, LVS activated carbon is a very interesting sorption material in terms of its commercialization and future practical use.

However, in the scientific literature at the time, there were no known methods of modifying LVS in order to obtain sorbents for the removal of anionic species from aqueous solutions, which opened a new field of research and wider possibilities of application of this economic and environmentally acceptable agro-waste material.

3. Eco-sorbents based on cationic LVS

Further research was carried out in the direction of the development of an original cationic sorbent based on lignocellulosic plant material, with the property of efficient sorption of anions from aqueous solutions (primarily phosphates and nitrates). In this sense, as a precursor for the synthesis of a new eco-sorbent, LVS biomass was investigated as a valuable agro-waste that had not been used for this purpose until then. However, the high mechanical stability of LVS biomass, which originates from the cross-linked lignocellulosic structure, does not favor the process of further modification in order to develop new or improve existing sorbents for the removal of anionic pollutants from contaminated waters. In this sense, the destruction of LVS biomass and partial delignification by the pretreatment process, before the synthesis of an adequate cationic type sorbent, is desirable. Bearing in mind that lignin is very stable in contact with water, weak bases and acids, as well as the fact that it is subject to chemical changes under the influence of elevated temperature, these properties must be taken into account when choosing the appropriate method and designing the biomass pretreatment procedure.

3.1 Pretreatment of lignocellulosic biomass

Pretreatment of lignocellulosic biomass involves partial delignification, and then depolymerization of hemicellulose, using appropriate reagents and optimal pretreatment conditions [33]. Delignification and depolymerization of hemicellulose increases biomass porosity, which during sorbent synthesis will enable better accessibility of agents for more efficient conversion of cellulose to the desired product. At the same time, the pretreatment enables the reduction of the required modifying agent amount, which represents an important economic and ecological effect. Pretreatment includes the separation of cellulose microfibrils from other biomass components (lignin, hemicellulose, pectin). This significantly increases the reaction surface, i.e. the number of reaction sites (primarily on cellulose) for the synthesis of the targeted sorbent. It should be kept in mind that the variability of the chemical composition and structure of the lignocellulosic material requires an appropriate pretreatment design. Therefore, numerous physical, chemical and biological methods have been developed for biomass pretreatment [3]. The choice of method should be in accordance with the expected outcome. In practical application, there are pretreatments of biomass: under mild alkaline conditions (for partial delignification while preserving cellulose), in an acidic solution (preferably for complete depolymerization of polysaccharides to monosaccharides), using organic solvents, by oxidative delignification (hydrogen peroxide or ozonolysis, preferably for lignin degradation), by biological methods (microbes) or microwave radiation [34].

In order to choose and design the pretreatment method, the physico-chemical characterization of LVS biomass was performed. The most abundant component in raw LVS biomass is lignin (41.90 ± 0.51%). The high content of holocellulose (57.80 ± 0.63%) indicates a significant presence of OH groups as active centers for further modification reactions. Holocellulose consists of cellulose (39.58 ± 0.42%) and hemicellulose (18.22 ± 0.14%). The protein content is extremely low (<0.1%), while volatile substances were not detected since the LVS was already naturally dried. The low ash content (0.28 ± 0.04%) mainly consists of bio-accumulated metals during plant growth, such as Ca, Zn, Cu and Mn (≈20 ppm). The biomass is characterized by a moisture content of 3.80 ± 0.14%, density of 0.46 ± 0.07 g/cm^3 and a bulk weight of 103.03 ± 0.87 kg/m^3. These data are consistent with the characteristics of other plant materials from the genus *Lagenaria* [20, 35].

In accordance with the established composition and purpose of the biomass, the alkaline pretreatment of LVS was chosen as the most suitable method. Considering that strongly alkaline reagents can lead to a significant loss of important polysaccharides, a green procedure of pretreatment with a green liquid as weakly alkaline reagent (1% Na_2CO_3 + 1% NaCl) was applied at a temperature of 23 ± 0.5°C, atmospheric pressure, for 2 hours with stirring (150 rpm). The resulting alkaline treated LVS biomass (LVSAT) was subjected to further characterization.

Namely, the most important parameters of pretreatment with green liquids are the biomass particle size (<10 mm, preferably about 1 mm), concentration of alkaline reagent, pH value, time and temperature [36]. In our case, for pretreatment purposes, LVS biomass was crushed and ground to a particle size in the range of 400–800 μm. The concentration of the applied reagent affects the yield of the solid phase of the biomass. During the pretreatment, a pH of around 9–10 is reached, which is significantly lower than in the case of treatment with alkaline cooking of biomass (pH 12–13). A lower pH value can effectively reduce the alkaline degradation and secondary reactions of polysaccharides during pretreatment. At the same time, a certain degree of swelling of cellulose fibers is achieved, which can be useful for further biomass modification. On the other hand, increasing the temperature shortens the pretreatment time, but affects the increased consumption of alkali and the formation of degradation products. Higher temperature contributes to the alkaline degradation of biomass, due to the breaking of ester bonds between lignin, hemicellulose and cellulose. Substituted uronic and acetyl groups on hemicellulose can be hydrolyzed to the form of uronic acid and acetic acid with more intensive alkaline degradation, which results in the formation of a larger proportion of acidic functional groups. Apart from the mentioned effects, alkaline pretreatment changes the macroscopic and microscopic properties of lignocellulosic biomass, including the size, structure and chemical composition of the particles.

In our case, the applied green procedure leads to a slight degradation of lignocellulosic biomass, while maintaining a higher cellulose content and increased delignification selectivity (**Figure 2**).

The advantage of applied alkaline pretreatment is the mild destruction of the lignostic part of the biomass, during which the ester bonds associated with hemicellulose xylan and lignin are hydrolyzed. Mild reaction conditions prevent the condensation of the released lignin fragments, which affects their greater solubility and their

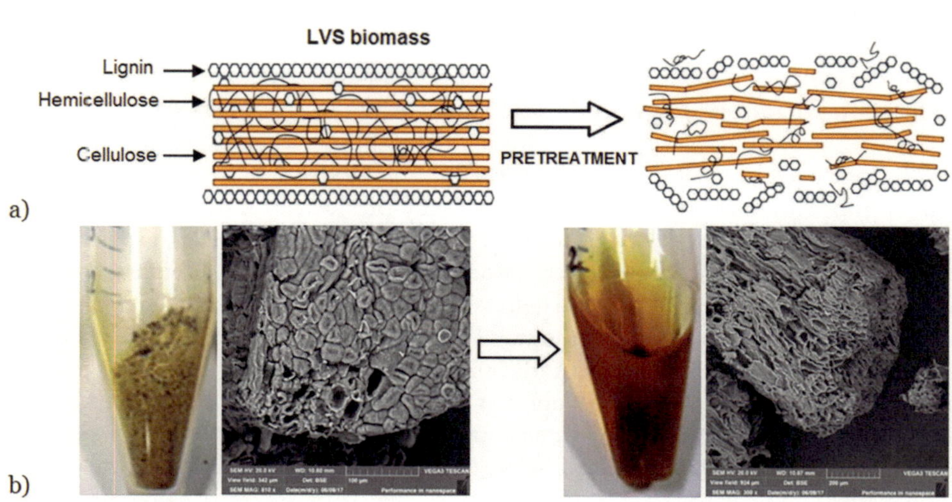

Figure 2.
Schematic presentation of LVS biomass pretreatment (a) and corresponding SEM micrographs (b) with images of raw LVS biomass before treatment (left) and brown-red filtrate after alkaline treatment of LVS biomass (right).

better removal from the biomass. The released hemicellulose dissolves in water and undergoes hydrolysis under the alkali action, which leads to its conversion into water-soluble monomers. Research has shown that this process removes uronic (through uronic acid) and acetyl groups (through acetic acid) present in hemicellulose. In this way, the availability of various chemical agents to the cellulose chain increases. These findings were confirmed by FTIR spectroscopic and SEM-EDS morphological analysis of biomass samples before (LVS) and after (LVSAT) the pretreatment procedure [37].

Under the specified pretreatment conditions, the mass loss of the initial LVS sample (15.2%), color change (from colorless to brown-red) and pH value of the filtrate after washing the LVSAT have been found. A significant decrease in the content of hemicellulose (71.4%) and partially lignin (5.2%) was registered, which directly affected the enrichment of biomass with cellulose (from 40 to 47%). Intense coloring of the reaction solution (from colorless to brown-red) during alkaline pretreatment of biomass originates from released pigments, hetero-saccharides and resulting degradation products (organic resin compounds, oxidized phenols and organic acids). Also, pectin (polymer of polygalacturonic acid) is released, whereby only that part of the pectin network where the glucosidic bond was previously broken can be separated. As a result of hemicellulose and pectin depolymerization, cinnamic, glucuronic and galacturonic acids are released. On that occasion, metal ions (such as Ca(II)) that kept the pectin structure more stable are also removed. The high pH value of the initial solution (about 10.1) during biomass pretreatment leads to breaking of ester bonds between lignin, hemicellulose and cellulose, as well as secondary reactions of polysaccharides (mainly released hemicellulose, which is much more sensitive to hydrolysis). Also, substituted uronic and acetyl groups on hemicellulose hydrolyze to the form of uronic and acetic acids, which results in the formation of more acidic substances. It is these organic acids that lead to a decrease in the pH value (from 10.1 to 8.6 during pretreatment). Used alkalis and released mineral substances are converted during the process into non-renewable salts. In order to remove undesirable salts and released oligosaccharide or monomer constituents, the biomass is filtered and subsequently washed with deionized water to a colorless filtrate (pH 6.8). After the applied pretreatment, it was determined that about 78% of the total polysaccharides are retained in the treated LVSAT biomass, which is important for further application as a sorbent synthesis precursor.

3.2 Functionalization of lignocellulosic biomass by cationic grafting

The increased interest in natural, renewable and biodegradable materials makes lignocellulosic biomass (primarily agricultural by-products or industrial waste biomass) important raw materials for the preparation of eco-sorbents, especially in the field of purification of polluted natural and wastewater [7]. Biomasses with cellulose as the main constituent are generally hydrophilic in nature. The presence of polar functional groups improves the stability of biomass in aqueous solutions, which is an advantage of these materials. Additionally, biomass as relatively porous materials is characterized by a high surface/volume ratio, which means that they have a highly reactive surface that can be easily functionalized [26]. The primary reactive sites for surface modification are mainly accessible hydroxyl and carboxyl groups. Various non-covalent and covalent modification approaches, such as radical polymerization and graft reactions, have been developed for the functionalization of biomass particles, as well as for improving their dispersibility in hydrophobic environments [38]. Alternatively, surface modification of lignocellulosic biomass can be achieved by sorption of the appropriate surfactant type (cationic or anionic), which is considered a non-covalent surface modification [12].

Surfactant-modified lignocellulosic biomasses have attracted a lot of attention as sorbents of various pollutants, considering that they are characterized by good performances, such as: low cost, biodegradability, suitable rheological properties, associative properties in water and surface-active properties that enable the sorption of various ionic pollutants matter [15, 16]. By the way, surfactants are defined as surface-active substances that greatly reduce the surface tension of water when used in very low concentrations [39]. However, one of the disadvantages of using surfactants for surface modification of the porous solid phase is that a large amount of surfactant is required due to the high specific surface area of the material. In this regard, some methods have been developed that are based on the use of a reduced amount of cationic surfactant [40]. For the surface modification of lignocellulosic biomass, the following cationic surfactants were mainly used: octadecyltrimethylammonium bromide (C18, OTAB), cetyltrimethylammonium bromide (C16, CTAB) or chloride (C16, CTAC), cetylpyridinium bromide (C16, CPB) or chloride (C16, CPC), tetradecyltrimethylammonium bromide (C14, TTAB) and dodecylpyridinium chloride (C12, DPC) [12]. For the purpose of producing sorbents, numerous biomasses or other substrates were modified: zeolite with CTAB for phosphate removal [41]; coconut kernel with CTAB to remove chromate, sulfate and thiocyanate [42]; yeast and lichen biomass with CTAB for chromate removal [43]; barley straw with CPC to remove food and mineral oils [44]; silkworm skeleton biomass with CTAB to remove anionic methyl orange dye [45]; cellulose fibers with CTAB [46]; wheat straw with CPB to remove anionic dyes [47]; tea waste with CTAB and CPB to remove the anionic Congo red dye [48].

In our case, prepared LVSAT biomass was modified with CTAC surfactant and investigated as a phosphate/nitrate sorbent in order to solve the problem of the eutrophication phenomenon [21]. Biomass modification was carried out at a temperature of 23 ± 0.5°C, during 8 h with mixing (150 rpm). To evaluate the sorption effect, two samples were prepared depending on the concentration of the surfactant solution: sample LVSAT-CTAC1 (0.9 × CMC) and sample LVSAT-CTAC2 (2 × CMC), where CMC is the critical micellar concentration of surfactant (1.28 mmol/dm^3).

In order to evaluate the added cationic groups in the surfactant-modified lignocellulosic biomass, elemental analysis (CHNS/O) was performed first. Changes in the content and type of elements of the initial LVSAT biomass and the obtained cationic product are shown in **Table 1**. As a function of the nitrogen (N) content, other characteristic parameters were also determined, such as: reaction efficiency (RE), degree of substitution (DS), amount of added cationic groups (N_{ad}) and product yield (Y_{pr}).

The elemental analysis of the samples determined that the value of CHO content in the biomass (H/C ratio 0.13) did not change significantly after the alkaline pretreatment. The presence of sulfur was not identified in the tested samples. Very similar results were observed in the case of research on other biomass [1, 20]. Significant changes in the nitrogen content were found in both samples of surfactant-modified biomass, which confirms the effectiveness of the cationization reaction. A higher N content in the LVSAT-CTAC2 sample was registered as a result of a more concentrated CTAC reagent solution (>CMC surfactant). The increase in N content was proportional to the concentrations of the solutions used. This fact indicates a possible mechanism of two-layer aggregation of surfactants on the biomass surface [42]. Based on the amount of added cationic groups in the biomass after modification, the theoretical ion exchange capacity of the LVAT-CTAC2 sorbent of 1.01 mmol N/g was calculated.

Samples	Elements (%)				N_{ad} (mmol/g)	RE (%)	DS	Yield (%)
	C	H	O	N				
LVS	45.46	5.98	48.55	0.01	—	—	—	—
LVSAT	45.51	5.84	48.65	0.01	—	—	—	84.4
LVSAT-CTAC1	52.33	6.67	40.20	0.79	0.56	97.39	0.11	97.4
LVSAT-CTAC2	60.09	6.95	31.52	1.43	1.01	78.91	0.25	97.9

$N_{ad} = (N_{product} - N_{biomass})\cdot 0.714$; $DS = 162\cdot N_{product}/(100\cdot 14 - M\cdot N_{product})$; $RE = 100\cdot N_{ad}/Q_{surfactant}$; and $Y_{pr} = 100\cdot(m_{product} - m_{biomas})$.

Table 1.
Elemental analysis and reaction efficiency of initial (LVS), alkali-treated (LVSAT) and its surfactant-modified (LVSAT-CTAC) lignocellulosic biomass.

Additionally, high values of reaction efficiency and product yield indicate maximum utilization of the CTAC reagent used. However, the small degree of substitution and the predicted two-layer mechanism suggest that the obtained data cannot be relevant for the sorption efficiency evaluation of anionic species from the solution [49]. This can be explained by the fact that at least half of the total amount of bound surfactants is occupied by interaction with biomass active centers, while the rest is available for the sorption process of anionic pollutants (**Figure 3**). Accordingly, for LVAT-CTAC2, we should expect an exchange capacity that is half the theoretically calculated one.

The effectiveness of the surfactant-modified LVSAT biomass as a cationic sorbent was evaluated by means of the sorption capacity for the tested anionic species (phosphates and nitrates). The mechanism of anions sorption involves ion exchange of weakly stable Cl^- counterions with ions of higher affinity ($H_2PO_4^-$ or NO_3^-) towards active ($-NR_4^+$) centers. As expected, low efficiency of anion sorption was registered with both sorbents (about 40% for phosphates and 22% for nitrates). Namely, the formation of two-layer surfactant aggregates increases the number of cationic added

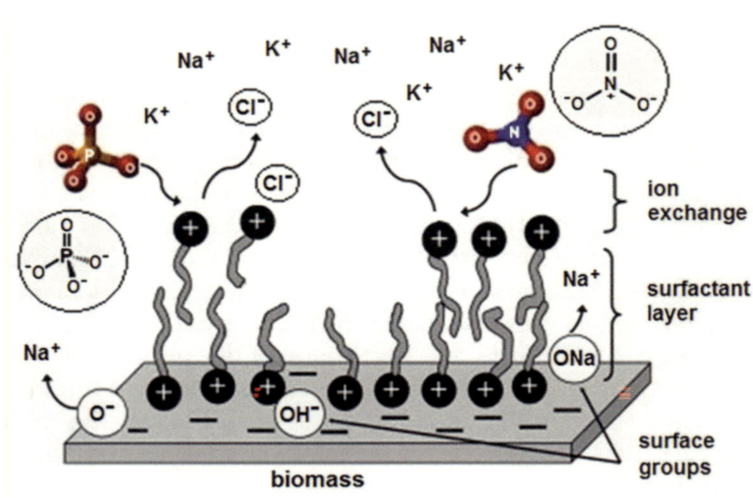

Figure 3.
Model of the CTAC surfactants organization on the LVSAT biomass surface and sorption of anionic species from aqueous solution (modified from Ref. [50]).

groups, but not the number of available active centers for interaction with the anions presents in the solution. In addition, the sorption efficiency is influenced by the pronounced steric effect of surfactant molecules. The layer of long hydrocarbon chains limits the access of anions to the less accessible active centers, distributed in the macroporous structure of the biomass. Also, the rheological properties of the solution due to the increase in surfactant concentration directly affect the difficult diffusion of anions through the solution during sorption, and thus lower sorption efficiency.

3.3 Synthesis of cationic biosorbents

Numerous cationization methods have been investigated for the chemical modification of lignocellulosic structures, mainly with secondary and tertiary amino reagents [2], as well as with quaternary ammonium reagents [11, 14]. The most commonly applied method of biomass cationization involves the cross-linking of epichlorohydrin (ECH) with appropriate amino reagents, such as trimethylamine (TMA), dimethylamine (DMA), diethylenetriamine (DETA), ethylenediamine (EDA), triethylamine (TEA), and others [1, 13]. Although the cationic sorbent synthesis method with ECH and TMA reagents is one of the most effective, with a high yield of the desired product as an anion exchanger, it should be noted that very toxic and harmful reagents are used, such as ECH, TMA and pyridine as a catalyst [51]. Therefore, less dangerous reactions for the modification of lignocellulosic biomass were recommended as an alternative [52]. Alternative methods for the synthesis of cationic sorbents include less toxic reagents, such as N-(3-chloro-2-hydroxypropyl)-trimethylammonium chloride (CHMAC) and glycidyl-trimethylammonium chloride (GTMAC). The application of these quaternary ammonium reagents is considered a safer and better option in terms of safer handling and environmental protection [14, 53].

In our case, following both synthesis procedures, two cationic biosorbents based on the prepared LVSAT precursor were synthesized in order to compare the efficiency of lignocellulosic biomass cationization, as well as their sorption characteristics. The LVSAT-TMA biosorbent was synthesized according to the first ECH-TMA method, while the less toxic CHMAC reagent was used for the synthesis of LVSAT-CHMAC biosorbent according to the second method [21].

According to the first ECH-TMA method, the reaction between ECH and cellulose is promoted by prior activation of primary and secondary OH groups of cellulose with an alkaline reagent (NaOH), resulting in alkali-cellulose with more reactive –ONa groups. The obtained hydroxy-cellulose ether is then cyclized in an alkaline medium using pyridine as a catalyst. Furthermore, epoxy-cellulose ether is used as an intermediate in the ammonolysis reaction. It is assumed that the ammonolysis reaction with TMA occurs after ring opening of the epoxy group, by condensation via the epoxy chloromethyl group in excess ECH [6]. ECH bound to the cellulose skeleton represents the active site of attack by amino reagents. Higher reaction efficiency and higher yield of the cationic product are achieved in the presence of N,N-dimethylformamide (DMF) as a medium and pyridine (PIR) as a catalyst. As a polar (hydrophilic) solvent with a high boiling point (152°C), DMF enables reactions that follow polar mechanisms to be carried out. On the other hand, as an aprotic solvent (it does not form hydrogen bonds, it solvates ions more weakly), it accelerates SN2 reactions given that nucleophiles are not strongly solvated. Although the reaction conditions (temperature and pressure) are not so drastic, in the case of interaction with alkali (NaOH) a part of DMF can be converted into dimethylamine, which increases the influence of the amino reagent on the cationic sorbent synthesis. Pyridine as a base has chemical properties similar to tertiary amines, so it is used as a catalyst for organic reactions. Apart from its presence in the solution accelerating the degree of wetting and causing the swelling of lignocellulosic biomass, pyridine is very easily protonated by available H^+ ions, forming a pyridinium cation. The use of DMF medium and pyridine as a catalyst increases the sensitivity of the epoxy ring to react with the –OH groups of the glucopyranose unit of cellulose. For this purpose, the Lewis acid leads to the appearance of a lowering of the activation energy for ring opening and allows the subsequent attachment of epichlorohydrin to the cellulose skeleton, as the target molecule for the initiation of the reaction [54]. The flow of successive chemical reactions of the cationic LVSAT-TMA biosorbent synthesis is shown in **Figure 4**.

In the second case, a procedure involving CHMAC reagent for the chemical modification of LVSAT precursor into LVSAT-CHMAC biosorbent was used. This sorbent synthesis procedure is simpler and more comprehensive compared to the previous ECH-TMA method (**Figure 5**). However, this synthesis process itself requires optimization of the reaction conditions which are peculiar to the nature of initial lignocellulosic biomass.

Many studies have dealt with the optimization of cationic sorbent synthesis using the CHMAC reagent, alternately changing one of the important parameters, such as temperature (60–100°C), time (2–24 h), medium (ethanol, water) and the quantitative NaOH/CHMAC ratio (0.5–2.0) [13, 14]. In our case, after a series of experiments in the function of preliminary phosphate and nitrate sorption, the following optimal conditions for the synthesis of LVSAT-CHMAC biosorbent were defined: aqueous medium, 5 M NaOH solution (25 mmol NaOH/g biomass) temperature 80°C, time 10 h, and the quantitative NaOH/CHMAC ratio of 1:1 (20 mmol CHMAC/g biomass) [21].

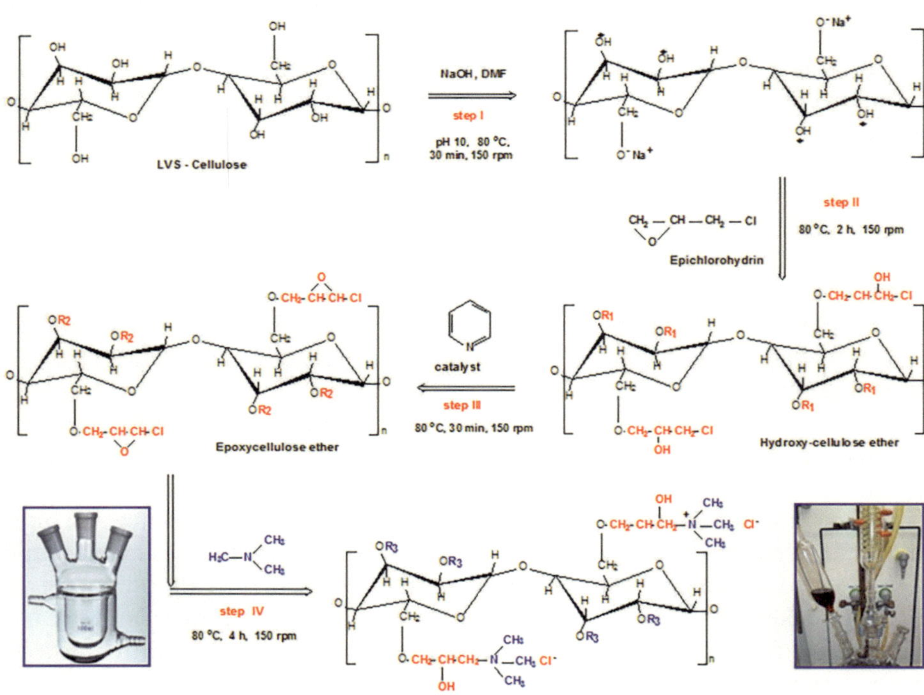

Figure 4.
Chemical reactions flow of the LVSAT-TMA biosorbent synthesis: Activation of LVS-cellulose (I), interaction of alkali-cellulose and ECH (II), conversion of hydroxy-cellulose to epoxy-cellulose ether (III), reaction of epoxy-cellulose ether with TMA to a cationic product (IV) [21].

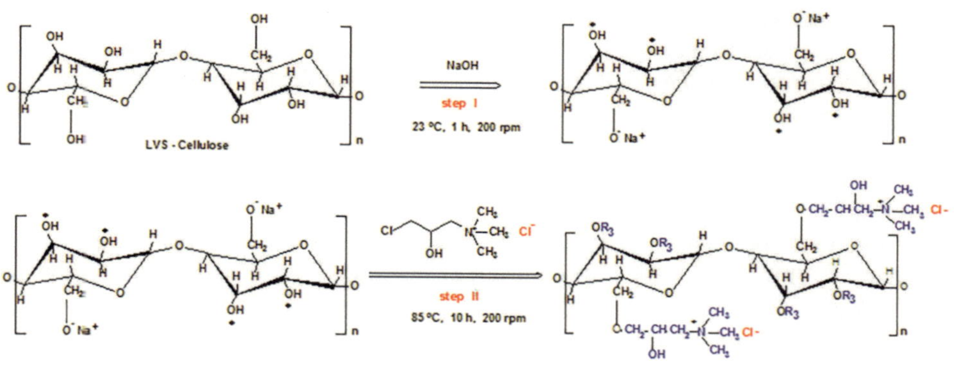

Figure 5.
Reaction path of the cationic LVSAT-CHMAC sorbent synthesis: Activation of LVS-cellulose monomer (I), reaction of alkali-cellulose with CHMAC reagent (II); potential active monomer groups are marked with asterisks [21].

The synthesized cationic biosorbents were characterized by elemental analysis and pH_{PZC}, including FTIR spectroscopy. Elemental (CHNS/O) analysis was applied in order to evaluate the added cationic groups in chemically modified lignocellulosic biomass. Changes in element content, reaction efficiency, and sorption properties of the synthesized biosorbents compared to LVSAT biomass are shown in **Table 2**.

Based on the elemental analysis of the tested samples, it was determined that 17.36 mgN/g in the form of a cationic $-N^+R_3$ group was bound on the biomass surface treated with the ECH-TMA reagent, which indicates the theoretical value of the LVAT-TMA sorbent ion exchange capacity of 1.24 mEq/g. In the case of the LVAT-CHMAC sorbent, it was observed that the quaternary ammonium CHMAC reagent contributes to a higher ion exchange capacity of the sorbent (1.39 mEq/g). Based on the DS and RE values, one gets the impression that substitution takes place on every fifth glucopyranose unit of the cellulose in the case of the TMA reagent, or on every third unit in the case of the CHMAC reagent. Assuming that RE is a property of the biomass rather than the amount of reagent, the initial concentration of the cationic reagent can be

Samples	Elements (%)				N_{ad} (mmol/g)	RE (%)	DS	Yield (%)
	C	H	O	N				
LVSAT	45.51	5.84	48.65	0.01	—	—	—	84.4
LVAT-TMA	47.67	6.34	44.25	1.74	1.24	4.86	0.22	94.6
LVAT-CHMAC	47.22	6.12	44.71	1.95	1.39	6.95	0.31	98.8
	Sorption efficiency							
Biosorbents	Phosphates (10 mgP/dm^3)				Nitrates (10 mgN/dm^3)			
	Q_e (mg/g)		%		Q_e (mg/g)		%	
LVSAT	0.05 ± 0.03		1.0		0.04 ± 0.02		0.8	
LVAT-TMA	3.94 ± 0.09		78.8		3.43 ± 0.14		68.6	
LVAT-CHMAC	4.41 ± 0.13		88.2		3.95 ± 0.10		79.0	

Table 2.
Elemental analysis and sorption characteristics of the synthesized cationic biosorbents compared to LVSAT biomass.

reduced. This is very important in a semi-industrial (scale-up) process, because it is one of the factors that determine the costs of the final product. This assumption can be explained by the fact that the reactive sites availability of lignocellulosic biomass is the limiting factor of reaction efficiency. In this sense, reaction barriers can be biomass density and the lignin amount. The bulk density of the biomass may be the result of the degree of lignification in this case, since the bulk density shows an inverse linear correspondence with the amount of added cationic groups [1]. Consequently, there may be diffuse limitations of the reaction between the cationic reagent and the reactive cellulose as a result of the bulk density (0.46 g/cm^3 in the case of LVS biomass). Similar findings that delignification of lignocellulosic materials can increase the efficiency of the quaternization reaction have been reported for other highly lignified biomasses [55].

The results of testing the change in surface charge of LVS biomass and LVSAT-CHMAC biosorbent as a function of pH value (from 3 to 11) are shown in **Figure 6**. The higher value of pH_{PZC} biosorbent (7.04) compared to pH_{PZC} of biomass (6.37) indicates the presence of cationic ($–N^+R_3$) functional groups that lead to an increase in the surface potential of the biomass. Also, the trend of decreasing potential of both biomass and sorbent with increasing pH (from 3 to 11) can be attributed to the influence of pH-dependent functional groups of biomass with negative zeta potential (OH group of cellulose and OCH_3 group of lignin skeleton). These groups show a higher negative charge with increasing pH, which results in a decrease in the positive charge of the tested samples [56]. Certainly, the analysis of the PZC confirmed the effectiveness of the biomass chemical modification by introducing quaternary ammonium groups that contribute to a more positive potential of the sorbent.

FTIR spectroscopic characterization of alkaline modified biomass (LVSAT) and synthesized biosorbent (LVSAT-CHMAC) was aimed at identifying structural changes of the precursor during synthesis and confirming the incorporation of quaternary ammonium groups into the biomass structure [37]. Corresponding segments of FTIR spectra in the areas of valence and deformation vibrations of functional groups with characteristic changes are shown in **Figure** 7. FTIR spectra of both analyzed LVS samples in the region 1800–500 cm^{-1} are typical for lignocellulosic biomass. In the FTIR spectrum of LVSAT biomass, the IR band at 1735 cm^{-1} originates from saturated ester C=O bonds of hemicellulose. The decrease in the intensity of this band indicates that during the alkaline pretreatment of biomass, the hydrolysis of glucosidic C-O-C bonds occurs, i.e. depolymerization of the released hemicellulose

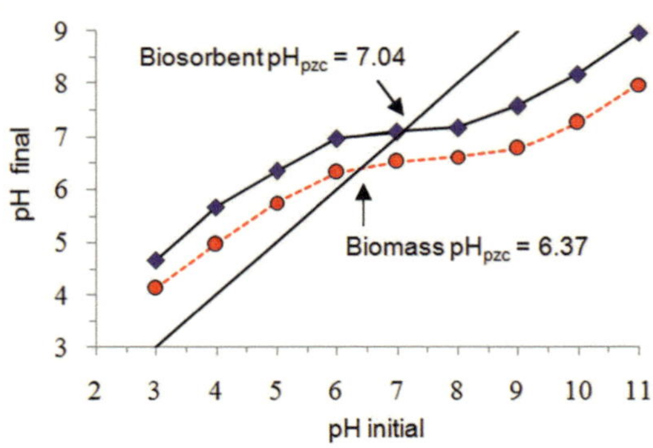

Figure 6.
The influence of pH on the surface charge of LVS biomass and LVSAT-CHMAC biosorbent.

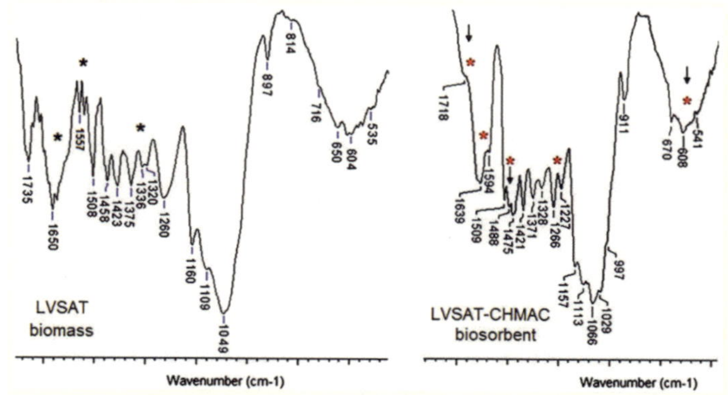

Figure 7.
FTIR spectra of LVSAT biomass and LVSAT-CHMAC biosorbent (significant changes are highlighted with asterisks).

and its partial removal from the biomass. In the spectral region 1700–1560 cm^{-1} there is a complex IR band, with centroid at about 1650 cm^{-1} and a shoulder at about 1600 cm^{-1}. This broad band indicates the overlap of the O-H vibrations of the crystal hydrates (water molecules in the lignocellulosic structure) with the present phenolic and carboxylic OH groups, as well as the skeletal C=C vibrations of the conjugated rings of lignin. The band at 1508 cm^{-1} originates from the skeletal vibration of the lignin aromatic ring. The doublet at around 1550 cm^{-1} indicates different types of aromatic C=C skeletal vibration, i.e. the presence of lignin fragments of different nature in the biomass. The subunits of lignin incorporated into the polymer can be guaiacyl, syringyl and p-hydroxyphenyl, which include methoxyl, acetyl and formyl groups.

This difference in composition has a great influence on delignification, as well as on biomass destruction. It is typical that guaiacyl units are more often cross-linked at the C_5 position of the aromatic ring, so that these cross C-C bonds make the delignification of biomass difficult, given that they cannot be hydrolyzed by either acid or base. In contrast, with syringyl units, this C_5 position is substituted, so it does not participate in further substitution reactions. The weak IR band at 1327 cm^{-1} represents the vibrations of the condensed syringyl G-ring, i.e. coniferyl alcohol substituted in the C_5 position. The splitting of this band into two peaks (1336 and 1320 cm^{-1}), as well as the appearance of a slightly more intense band at 1260 cm^{-1}, can be explained by the increased number of C-OH bonds due to the presence of ferulic acid in the lignin polymer. The IR band at about 1460 cm^{-1} originates from the deformation C-H vibrations. IR bands at 1423 cm^{-1} (O-H bending vibrations) and 1260 cm^{-1} (vibrations of cellulose ester group) are correlated with the crystalline structure of cellulose. Small changes in the intensity of these IR bands may indicate a partial transition from the crystalline to the amorphous form of cellulose in the treated LVSAT sample, which favors a more efficient synthesis of sorbents based on such a precursor. The broad IR band in the region 1100–900 cm^{-1} indicates a significant overlap of C-OH and C-C-O vibrations. The IR band at 897 cm^{-1} and the shoulder at 990 cm^{-1} originate from the C-H bending vibrations, which are characteristic of the anomeric β-glucosidic (C-O-CH) bonds of cellulose. These bands remain unchanged during the treatment of biomass, which indicates the structural stability of the cellulose chain under the applied reaction conditions. The spectral region below 800 cm^{-1} corresponds to the deformation vibrations of the C-H and O-H partners [57].

In the FTIR spectrum of LVSAT-CHMAC biosorbent (**Figure 7**), IR bands at 1030, 1066, 1113 and 1157 cm^{-1} typical for O-H, C-OH, C-O and C-H vibrations of OH and CH_2 groups of cellulose glucopyranose units were identified [57]. The IR bands at higher wavenumbers (1266, 1328, 1371, 1421, 1475, 1509 and 1594 cm^{-1}) correspond to the vibrations of OH, CH_2, O-CH_3, CO and C=C functional groups typical of lignin [4]. The weakened IR band at 1718 cm^{-1} originates from the C=O vibration of the remaining hemicellulose (trapped in the biomass structure). The assigned absorption bands were used to detect structural changes during the chemical treatment of LVSAT biomass.

Based on spectral-structure correlation and comparative analysis of FTIR spectra shown in **Figure 7**, it is evident that chemical modification of biomass with CHMAC reagent induces significant changes in the number, position and intensity of some IR bands. This confirms the decrease in the intensity of the IR bands originating from the vibrations of all types of OH groups (1266, 1030 and 608 cm^{-1}), the hemicellulose C=O group (1718 cm^{-1}), and–CH groups of the glucosidic bonds (911 cm^{-1}). Also, there is a shift in the positions of the IR bands characteristic of cellulose (from 1260 to 1266 cm^{-1}, from 1040 to 1066 cm^{-1} and from 897 to 911 cm^{-1}) and hemicellulose (from 1738 to 1718 cm^{-1}). These changes clearly indicate a partial disruption of the biomass lignocellulosic structure and the destruction of the hemicellulose glycosidic structure. The decrease in the intensity of the band at 600 cm^{-1} (from C=C), with a simultaneous shift to 1594 cm^{-1}, as well as the change in the position of the band from 1462 to 1475 cm^{-1} originating from lignin -OCH_3 groups, confirm the partial disruption of the lignin structure. Additionally, the IR band at 1260 cm_{-1} (from the C-OH group) splits into two new bands (at 1266 and 1227 cm^{-1}), indicating the appearance of OH groups of different origin, probably from the glucopyranose unit of cellulose and the trimethylammonium- hydroxypropyl agent. A similar phenomenon was observed in the region of valence C-O vibrations, where the band at about 1040 cm^{-1} doubles into two bands (1066 and 1029 cm^{-1}), which confirms the previous statement.

The most important proof of the biosorbent synthesis success is the appearance of a new band of lower intensity (due to overlapping) at 1490 cm^{-1}, which corresponds to asymmetric C-H vibrations of the introduced quaternary $–N^+(CH_3)_3$ functional group. The partner of this IR band (from the asymmetric C-N bond) is located at about 980 cm^{-1}, suggesting the presence of new $–CH_3$ groups. All these observations clearly indicate the incorporation of the cationic $–N^+R_3$ group on the LVSAT-CHMAC biosorbent surface during biomass modification [4, 6, 37].

3.4 Biosorbent efficiency and mechanism of sorption process

The efficiency of the synthesized LVSAT-CHMAC biosorbent was tested through the processes of removing phosphate and nitrate anions from aqueous solutions, which appear more and more often in natural and especially in wastewater. After a series of sorption experiments in batch mode, an analysis of the process parameters influence on the biosorbent sorption capacity was performed. The following optimal parameters were determined for the sorption of phosphate ions: sorbent dose (2 g/dm^3), sorbent/sorbate contact time (30–40 min), medium pH (6.0 ± 0.2), temperature (20.0 ± 0.5°C), and mixing speed (150 rpm). Based on the conducted tests (**Figure 8**), the sorption capacity of the LVSAT-CHMAC biosorbent for phosphate ions was determined to be 17.8 mg P/g at optimal conditions.

In the case of nitrate sorption, the following optimal parameters are defined: sorbent dose (2 g/dm^3), sorbent/sorbate contact time (20–30 min), medium pH

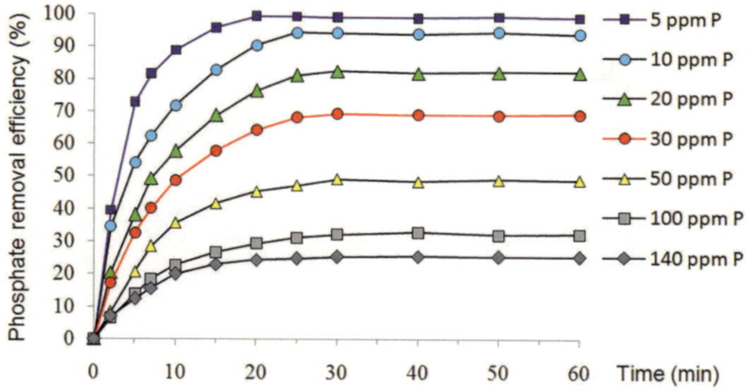

Figure 8.
Phosphate removal efficiency from aqueous solutions of different initial concentrations using LVSAT-CHMAC biosorbent as a function of sorption time, according to optimal sorption parameters.

(6.5 ± 0.2), temperature (20.0 ± 0.5°C), and mixing speed (150 rpm). Based on the conducted tests (**Figure 9**), the sorption capacity of the LVSAT-CHMAC biosorbent for nitrate ions was determined to be 15.7 mg N/g at optimal conditions.

In order to examine the qualitative and semi-quantitative elemental composition of both biomass and biosorbent surface, the Energy-dispersive X-ray spectroscopy (EDS) was used. This method combined with SEM microscopy, in addition to elemental composition, also provided data on the elements location on the analyzed biosorbent surface after phosphate and nitrate sorption. The results of SEM-EDS analysis of the samples before and after anion sorption are presented in **Figure 10**. It should be noted here that the Cr signals originate from the electroconductive thin layer of chromium used during the preparation of samples for analysis. SEM-EDS analysis of the examined samples provided direct evidence of the incorporation of quaternary ammonium groups ($NR_4^+Cl^-$) on the biosorbent surface after the biomass chemical modification, by detecting new N signals. The sorption efficiency of phosphate and nitrate anions on the LVSAT-CHMAC biosorbent surface was confirmed by the detection of new P and K signals (from $K^+H_2PO_4^-$), as well as more intense N signals (from NO_3^-), respectively. Characteristic changes of the Cl signal in the spectrum of the biosorbent, whose intensity significantly decreases after sorption, indicated

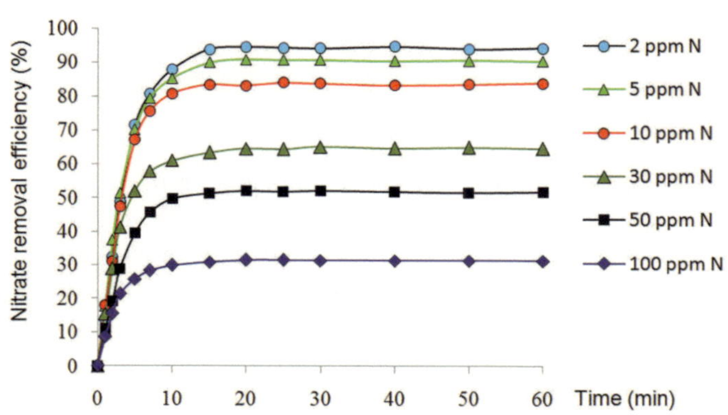

Figure 9.
Nitrate removal efficiency from aqueous solutions of different initial concentrations using LVSAT-CHMAC biosorbent as a function of sorption time, according to optimal sorption parameters.

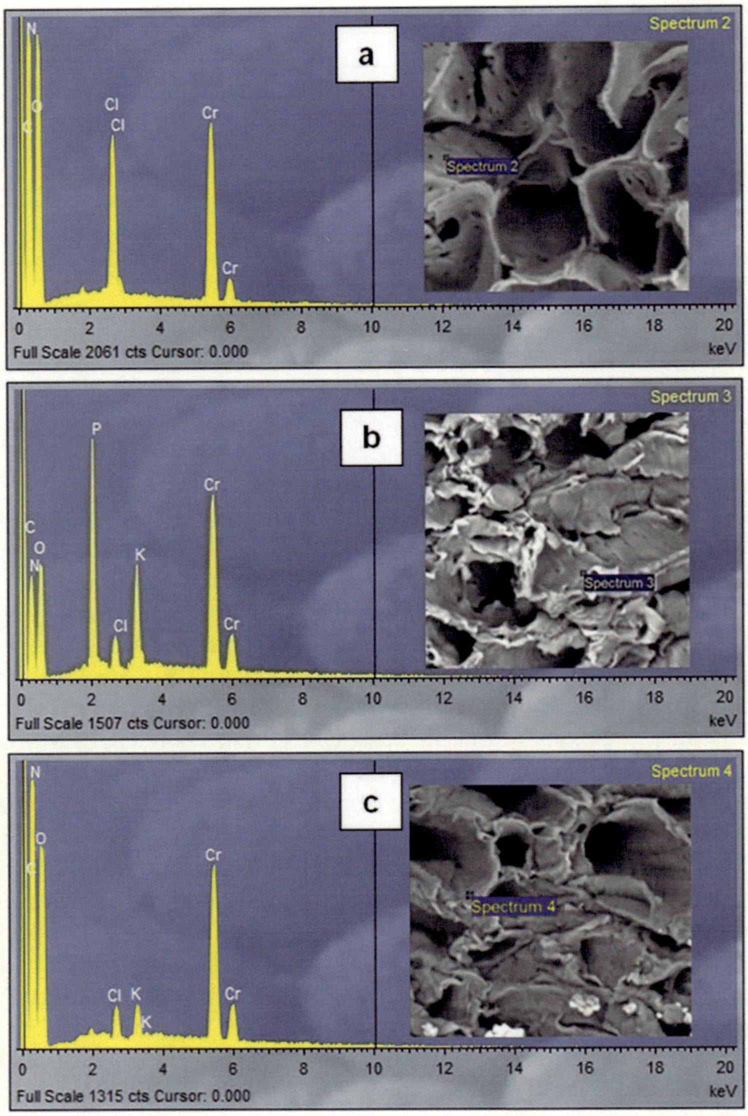

Figure 10.
EDS spectrum (left) and corresponding SEM micrograph (right) of LVSAT-CHMAC biosorbent before sorption (a), after phosphate sorption (b) and after nitrate sorption (c).

a possible ion exchange mechanism of chloride ions with phosphate or nitrate ions in this process. Other studies also point to ion exchange as the dominant sorption mechanism of these anionic species by different agricultural wastes [13].

Phosphate and nitrate sorption is best described by a pseudo-first-order nonlinear kinetic model (R^2 in the range 0.990–0.998). The complex nature of the sorption process, based on both surface sorption and diffusion within the biosorbent particles, was confirmed by the Weber-Morris model. This model indicated the dominant effect of the boundary layer, which has a direct influence in limiting the overall speed of the sorption process [58–60]. Phosphate sorption equilibrium is subject to laws of Freundlich and Sips isotherms, which indicates the complex nature

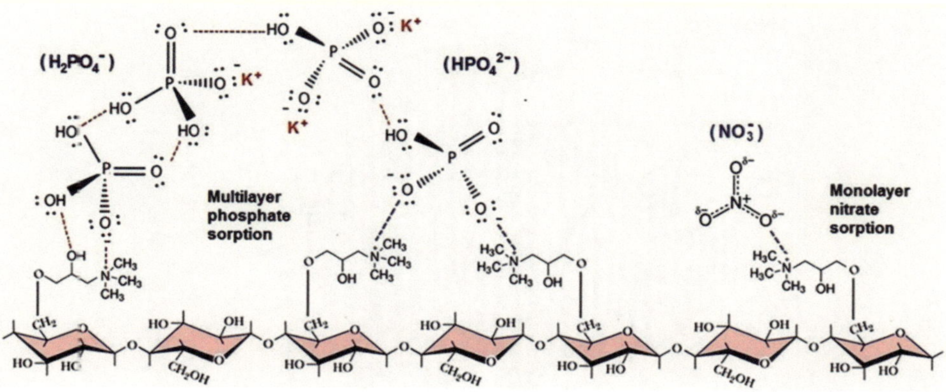

Figure 11.
Structural model of sorption of dominant phosphate ($H_2PO_4^-$ and HPO_4^{2-}) and nitrate (NO_3^-) anions from aqueous solutions using LVAT-CHMAC biosorbent.

of the sorption process (physical sorption and ion exchange). The mutual interaction of sorbed anions represents the accompanying mechanism, which leads to the formation of a multimolecular layer of phosphate on the biosorbent surface. On the other hand, the equilibrium of nitrate sorption is subject to the laws of the Sips and Langmir isotherms, which means that nitrate sorption takes place in a monomolecular layer, on an energetically uniform surface of the biosorbent with a finite number of active binding centers, without mutual interaction and trans-migration of ions on the biosorbent surface [58–60]. The corresponding model of phosphate and nitrate sorption on the biosorbent surface is shown in **Figure 11**. Thermodynamic studies of the investigated anions sorption indicated a spontaneous exothermic process in the temperature range of 20–40°C. At higher temperatures, the process of anions desorption from the biosorbent surface is favored. This fact and a slight decrease in the disorder of the system at the biosorbent/solution interface, confirm ion exchange as the most likely sorption mechanism [58–60].

4. Experimental design and sorption process optimization

The fact is that the efficiency of pollutant removal from contaminated aqueous solutions, as well as the sorption capacity of the biosorbent, depends on the choice and way of varying numerous reaction conditions during the sorption process (pH, temperature, sorbent dose, solution concentration, mixing speed, contact time, etc.). In this sense, in order to determine the key process parameters and their optimization, it is necessary to conduct a very large number of experiments (**Figure 12**). At the same time, it should be taken into account that in each series of experiments only one parameter varies, while the others are kept constant. It requires a lot of time, patience and persistence, and it also creates additional costs. Despite this, the conventional approach to such research (including kinetic, equilibrium and thermodynamic studies) leaves a lot of experimental gaps important for predicting optimal sorption parameters.

Therefore, in order to optimize the sorption process and predict optimal factors (experimental conditions), it is desirable to apply the experimental design methodology. This study involves the integration of the OVAT methodology (One-Variable-At-a-Time) with a factorial design (Design of Experiments - DoE). The advantage of

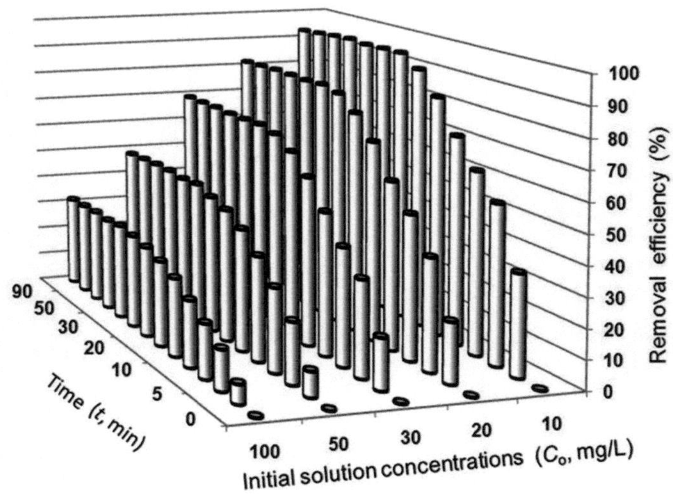

Figure 12.
Example of experiment series on phosphate removal efficiency as a function of two variables (initial solution concentration and sorbent-sorbate contact time).

the DoE methodology is a significantly smaller number of experiments that should be conducted with the aim of evaluating the sorption capacity of the sorbent, the interaction effects of independent variables on the sorption efficiency, as well as the clarification of the sorption mechanism.

In our study on the phosphate anions sorption using LVAT-CHMAC biosorbent, the critical factors of the sorption process were determined by the OVAT method [61]. In order to optimize and evaluate the factors interaction effects, the Central Composite Design (CCD) was used within the Response Surface Methodology (RSM) analysis. Using the OVAT approach, the four most important input variables were determined (A - initial concentration of the phosphate solution, B - pH value, C – temperature, and D - time), as well as their individual trends that significantly affect the phosphate sorption process. Within the planning of experiment, a CCD matrix with the specified 4 factors on three levels (−1, 0, +1) was randomly generated using adequate statistical software (in our case, JMP-Pro 16 by SAS). The estimated values of critical variables represent the zero coded values for the CCD design. Based on the set data, the plan included 26 experiments, where eight axial points will be used to evaluate the square terms, 16 factor points to evaluate the main effects and two-way interactions, and two central points to evaluate the model adequacy. The outcome of the combined variables influence is the response with predicted values of sorbed phosphate amount.

The RSM-CCD design methodology was useful in forming a complete picture of the phosphate sorption process by revealing significant effects of variables and determining the effects of factor interactions on the response. In order to predict the sorption capacity, a second-order polynomial model was developed and validated (Eq. (1)). This model reveals the influence of the combined process factors (A, B and C) on the response (Y). Initial phosphate concentration (A) was identified as the main process factor with a positive impact (synergism). A negative influence of the temperature (C) was found, while the sorbent-sorbate contact time (D) was not a statistically significant factor (given that the equilibrium of the process is established after 40 minutes).

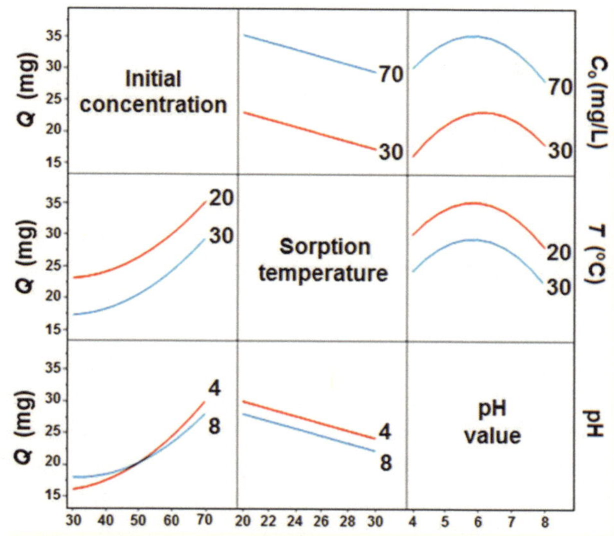

Figure 13.
Profiles of the factors interaction effect on the response Q (sorbed phosphate amount in mg).

$$Y = 23.52 + 5.98\,A - 2.88\,C - 0.97\,AB + 2.74\,A^2 - 6.13\,B^2 \qquad (1)$$

Also, the interaction between initial phosphate concentration and pH value (AB) was found to have a significant effect on the predicted response (**Figure 13**). In addition to identifying and testing significance of the main process factors, the RSM metodology indicated the importance of pH value for the phosphate removal from aqueous solution. Namely, different forms of phosphate anions ($H_2PO_4^-$, HPO_4^{2-}, and PO_4^{3-}) are represented in aqueous solutions. As a critical factor, pH can affect their sorption differently, primarily depending on the LVAT-CHMAC biosorbent pH_{PZC} value. Thus, although increased anion sorption is expected at pH < pH_{PZC} (due to the positively charged surface of the biosorbent), it was observed that the sorption efficiency decreases with decreasing pH. The decrease in sorption efficiency ($PO_4^{3-} > HPO_4^{2-} > H_2PO_4^- > H_3PO_4$) can be explained by the pronounced protonation effect of anions to the H_3PO_4 molecules in more acidic solutions. The maximum phosphate sorption efficiency is achieved at pH between 5 and 7, by the ion exchange mechanism of the dominant HPO_4^{2-} and $H_2PO_4^-$ forms. In alkaline medium (pH > pH_{PZC}), the sorption of PO_4^{3-} form decreases due to competition with OH^- ions, which indicates the mechanism of electrostatic interaction between the sorbate and the biosorbent surface.

In accordance with the applied experiment design methodology, the sorption capacity average value of 17.6 mgP/g at optimal process parameters (C_o = 70 mg/L, t = 40 min, pH = 5.8, and T = 20°C) was predicted (with 95% certainty). The close match between the predicted and experimental value (17.8 mgP/g) of the sorption capacity confirmed the applicability of the developed regression model.

5. Biosorbent valorization

Investigations of the potential interference of coexisting anions in the solution on the sorption capacity of LVAT-CHMAC biosorbents showed that the presence of nitrates and sulfates leads to a decrease in phosphate sorption capacity by about 35%, while chlorides did not show a significant effect. Phosphate sorption interference follows the order: NO_3^- > SO_4^{2-} > Cl^-. Phosphates were more competitive than nitrates in the sorption process of anions from the binary mixture [62].

The possibility of LVAT-CHMAC biosorbent regeneration was monitored through five consecutive cycles of the sorption/desorption process, as a function of the anions initial concentration (5–100 mg/dm^3). In the first cycle, desorption degree of anions from 99 to 95% was achieved. This indicates a reversible process, as well as the fact that there is no strong interaction between the biosorbent and the sorbed anions. The biosorbent effectiveness was also confirmed in the fifth cycle, when a desorption rate of about 50% was achieved.

Our study indicated that the sorption capacities of LVAT-CHMAC biosorbent for phosphates (0.58 mmol P/g) and nitrates (1.18 mmol N/g) are comparable to the results of most other cationic sorbents tested for anions removal under similar conditions [21]. Thus, the phosphate removal ability by modified LVS biomass was higher or very similar to walnut shell, almond shell, peanut shell, rice husk, corn cob, oak chip, palm kernel, wheat straw (0.45–0.71 mmol/g), even in the case of some commercial exchange resins (such as quaternary ammonium cellulose QA52). In the case of nitrates, a higher sorption capacity of LVS biosorbent was determined in relation to cationically modified biomasses such as coconut husk, wheat or rice straw, coconut fibers, sugar cane and various sawdust, even compared to some mineral clays such as halloysite and bentonite.

From the techno-economic aspect, the production cost analysis of any adsorbent is an important factor in the profitability of its practical application in the water purification process. Therefore, the production total cost of the cationic LVS biosorbent was estimated. Production costs are influenced by numerous factors such as: availability of raw material (LVS biomass), processing and treatment requirements (synthesis), as well as reuse of biosorbent. The total cost includes the price of each individual technological operation (rinsing, drying, grinding, sieving, activation, neutralization, synthesis, heating, and mixing). Here, it is taken into account that the bottle gourd shell was collected as agricultural waste (free of cost) and dried naturally (which was not included in the processing cost). Based on the reagents costs and the price of electricity, the approximate total price for the biosorbent production is estimated at 11.5 EUR/kg. This sum is comparable to the production costs of other sorbents based on similarly modified lignocellulosic biomasses (10–25 USD/kg). Therefore, LVAT-CHMAC biosorbent can be considered an economical resource in water treatment technology [60].

In addition to the established properties (economical production, efficient sorption of anionic species, possibility of regeneration), a significant advantage of using LVAT-CHMAC biosorbent is ecological utilization at the end of the life cycle. As a biodegradable material, it is not harmful to the environment. On the other hand, the multiple sorptions-desorption process makes it possible to obtain a concentrated phosphate-nitrate solution or to accumulate these anions in the biosorbent. This solution, as well as the used biosorbent enriched with nutrient elements available for plants (P, K, N, and minerals) can be used as compost for feeding plants or making artificial fertilizers for acidic soils in dry climate areas. In this way, the ecological problem of disposing of the used sorbent or the removed sorbate is also solved.

6. Conclusion

The study indicates that sorbents with significantly improved sorption characteristics can be obtained by simple and accessible procedures of chemical-thermal modification of bottle gourd shell as a cheap and available agro-waste material. The LVS sorbent in form of activated carbon has a high potential for removing various cationic pollutants from aqueous solutions, primarily toxic metals, as well as organic non-polar and weakly polar compounds, in a wide range of sorption parameters (initial concentration of pollutants, pH value and temperature).

Conducted research validates the superiority of LVS biomass as a precursor for cationic modification with quaternary ammonium reagents compared to other agricultural by-products. The sorption efficiency of the cationic LVS biosorbent is comparable to more expensive commercial anion exchange resins tested for the removal of anionic species from wastewater, mainly phosphate and nitrate as the cause of eutrophication. In accordance with the technological, economic and ecological requirements of contaminated water purification, the following process parameters are recommended for effective anions removal using LVS biosorbent: temperature 18–22°C, slightly acidic solution (pH 5–7), initial concentration of anion solution 30–50 mg/dm^3 (removal rate 90–60%, respectively), sorbent dose 2 g/dm^3 for low level polluted or 5 g/dm^3 for more contaminated solutions, contact time 30–40 min.

High sorption capacities, rapid achievement of sorption equilibrium, relatively simple and low cost production, as well as reuse (by multiple sorption/desorption process) indicated that LVS sorbents can be competitive on the market of materials applicable in water purification technology. In order to solve the problem of waste disposal, although it is biodegradable and harmless to the environment, the biosorbent and preconcentrated solution after desorption can be utilize to fertilizer or compost produce.

Acknowledgements

This research was supported by the Ministry of Science, Tehnological Development and Innovation of the Republic of Serbia within the Program of scientific work (No. 451-03-65/2024-03/ 200133; 451-03-65/2024-03/200124; 451-03-66/2024-03/200124).

Conflict of interest

The authors declare no conflict of interest.

Author details

Goran Nikolić[1*], Dragana Marković Nikolić[2], Aleksandar Bojić[3], Danijela Bojić[3], Ljubiša Nikolić[1], Ljiljana Stanojević[1], Miloš Durmišević[1], Nataša Simonović[1] and Miloš Kostić[3]

1 Faculty of Technology, University of Niš, Niš, Serbia

2 Department of Technological Art Studies, Academy Southern Serbia, Leskovac, Serbia

3 Faculty of Sciences and Mathematics, Department of Chemistry, University of Niš, Niš, Serbia

*Address all correspondence to: goranchem_yu@yahoo.com; gnikolic@tf.ni.ac.rs

References

[1] Wartelle LH, Marshall WE. Quaternized agricultural by-products as anion exchange resins. Journal of Environmental Management. 2006;**78**(2):157-162. DOI: 10.1016/j.jenvman.2004.12.008

[2] Orlando US, Baes AU, Nishijima W, Okada M. A new procedure to produce lignocellulosic anion exchangers from agricultural waste materials. Bioresource Technology. 2002;**83**(3):195-198. DOI: 10.1016/S0960-8524(01)00220-6

[3] Alvira P, Tomas-Pejo E, Ballesteros M, Negro MJ. Pretreatment technologies for an efficient bioethanol production process based on enzymatic hydrolysis: A review. Bioresource Technology. 2010;**101**(13):4851-4861. DOI: 10.1016/j.biortech.2009.11.093

[4] Krishnan KA, Haridas A. Removal of phosphate from aqueous solutions and sewage using natural and surface modified coir pith. Journal of Hazardous Materials. 2008;**152**(2):527-535. DOI: 10.1016/j.jhazmat.2007.07.015

[5] Kumar P, Sudha S, Chand S, Srivastava VC. Phosphate removal from aqueous solution using coir-pith activated carbon. Separation Science and Technology. 2010;**45**(10):1463-1470. DOI: 10.1080/01496395.2010.485604

[6] Xu X, Gao B, Wang W, Yue Q, Wang Y, Ni S. Adsorption of phosphate from aqueous solutions onto modified wheat residue: Characteristics, kinetic and column studies. Colloids and Surfaces B: Biointerfaces. 2009;**70**(1):46-52. DOI: 10.1016/j.colsurfb.2008.12.006

[7] Huang X, Liu H, Shang S, Rao X, Song J. Preparation and characterization of polymeric surfactants based on epoxidized soybean oil grafted hydroxyethyl cellulose. Journal of Agricultural and Food Chemistry. 2015;**63**(41):9062-9068. DOI: 10.1021/acs.jafc.5b03765

[8] Xu X, Gao B-Y, Yue Q-Y, Zhong Q-Q. Preparation of agricultural by-product based anion exchanger and its utilization for nitrate and phosphate removal. Bioresource Technology. 2010;**101**(22):8558-8564. DOI: 10.1016/j.biortech.2010.06.060

[9] Haugen KS, Semmens MJ, Novak PJ. A novel in-situ technology for the treatment of nitrate contaminated groundwater. Water Research. 2002;**36**(14):3497-3506. DOI: 10.1016/S0043-1354(02)00043-X

[10] Samatya S, Kabay N, Yuksel U, Arda M, Yuksel M. Removal of nitrate from aqueous solution by nitrate selective ion exchange resins. Reactive and Functional Polymers. 2006;**66**(11):1206-1214. DOI: 10.1016/j.reactfunctpolym.2006.03.009

[11] Xu X, Gao B, Yue Q, Zhan X. Preparation, characterization of wheat residue based anion exchangers and its utilization for the phosphate removal from aqueous solution. Carbohydrate Polymers. 2010;**82**(4):1212-1218. DOI: 10.1016/j.carbpol.2010.06.053

[12] Alila S, Boufi S, Belgacem MN, Beneventi D. Adsorption of a cationic surfactant onto cellulosic fibers: Surface charge effects. Langmuir. 2005;**21**(18):8106-8113. DOI: 10.1021/la050367n

[13] Nguyen TAH, Ngo HH, Guo W, Nguyen TV. Phosphorous removal from aqueous solutions by agricultural

by-products: A critical review. Journal of Water Sustainability. 2012;**2**(3):193-207

[14] Keränen A, Leiviskä T, Gao B-Y, Hormi O, Tanskanen J. Preparation of novel anion exchangers from pine sawdust and bark, spruce bark, birch bark and peat for the removal of nitrate. Chemical Engineering Science. 2013;**98**:59-68. DOI: 10.1016/j.ces.2013.05.007

[15] Delforno TP, Moura AGL, Okada DY, Varesche MBA. Effect of biomass adaptation to the degradation of anionic surfactants in laundry wastewater using EGSB reactors. Bioresource Technology. 2014;**154**:114-121. DOI: 10.1016/j.biortech.2013.11.102

[16] Kaboorani A, Riedl B. Surface modification of cellulose nanocrystals (CNC) by a cationic surfactant. Industrial Crops and Products. 2015;**65**:45-55. DOI: 10.1016/j.indcrop.2014.11.027

[17] Rahman AS. Bottle gourd (*Lagenaria siceraria*): A vegetable for good health. Natural Product Radiance. 2003;**2**(5):249-256

[18] Prajapati RP, Kalariya M, Parmar SK, Sheth NR. Phytochemical and pharmacological review of *Lagenaria sicereria*. Journal of Ayurveda and Integrative Medicine. 2010;**1**(4):266-272. DOI: 10.4103/0975-9476.74431

[19] Mohan R, Birari R, Karmase A, Jagtap S, Bhutani KK. Antioxidant activity of a new phenolic glycoside from *Lagenaria siceraria* stand. fruits. Food Chemistry. 2012;**132**(1):244-251. DOI: 10.1016/j.foodchem.2011.10.063

[20] Mitic-Stojanovic D-L, Zarubica A, Purenovic M, Bojic D, Andjelkovic T, Bojic A. Biosorptive removal of Pb^{2+}, Cd^{2+} and Zn^{2+} ions from water by *Lagenaria vulgaris* shell. Water SA. 2011;**37**(3):303-312. DOI: 10.4314/wsa.v37i3.68481

[21] Marković DZ. Synthesis of cationic lignocellulosic sorbents and application for the removal of anionic pollutants from water [thesis]. University of Niš, Faculty of Sciences and Mathematics; 2018

[22] Ljupković R, Mitrović J, Radović M, Kostić M, Bojić D, Mitić-Stojanović D-L, Bojić A. Removal Cu(II) ions from water using sulphuric acid treated *Lagenaria vulgaris* shell (Cucurbitaceae). Biologica Nyssana. 2011;**2**(2):85-89

[23] Stanković M, Krstić N, Slipper I, Mitrović J, Radović M, Bojić D, et al. Chemically modified *Lagenaria vulgaris* as a biosorbent for the removal of Cu(II) from water. Australian Journal of Chemistry. 2012;**66**(2):227-236. DOI: 10.1071/CH12422

[24] Kostić M, Radović M, Mitrović J, Antonijević M, Bojić D, Petrović M, et al. Using xanthated *Lagenaria vulgaris* shell biosorbent for removal of Pb(II) ions from wastewater. Journal of the Iranian Chemical Society. 2014;**11**(2):565-578. DOI: 10.1007/s13738-013-0326-1

[25] Bojić D, Nikolić G, Mitrović J, Radović M, Petrović M, Marković D, et al. Kinetic, equilibrium and thermodynamic studies of Ni(II) ions sorption on sulfuric acid treated *Lagenaria vulgaris* shell. Chemical Industry and Chemical Engineering Quarterly. 2016;**22**(3):235-247. DOI: 10.2298/CICEQ150318037B

[26] Bojić D, Ranđelović M, Zarubica A, Mitrović J, Radović M, Purenović M, et al. Comparison of new biosorbents based on chemically modified *Lagenaria vulgaris* shell. Desalination and Water Treatment. 2013;**51**(34-36):6871-6881. DOI: 10.1080/19443994.2013.771287

[27] Bojić D. Development and application of eco-sorbents based on modified lignocellulosic biomass [thesis]. University of Niš, Faculty of Technology Leskovac; 2016

[28] Lagenaria Siceraria. Natural Resources Conservation Service, Plants Database. USDA. Available from: https://plants.usda.gov/home/plantProfile?symbol=LASI; 2023 [Accessed: December 20, 2023]

[29] Mitić D-L. Removal of Heavy Metals from Water Using Biosorbent Based on *Lagenaria vulgaris* [Thesis]. University of Niš, Faculty of Sciences and Mathematics; 2012

[30] Bojić D, Momčilović M, Milenković D, Mitrović J, Banković P, Velinov N, et al. Characterisation of a low cost *Lagenaria vulgaris* based carbon for ranitidine removal from aqueous solutions. Arabian Journal of Chemistry. 2017;**10**(7):956-964. DOI: 10.1016/j.arabjc.2014.12.018

[31] Kostić M, Radović M, Mitrović J, Bojić D, Milenković D, Bojić A. Application of new biosorbent based on chemicaly modified *Lagenaria vulgaris* shell for the removal of copper(II) from aqueous solutions: Effects of operational parameters. Hemijska Industrija. 2013;**67**(4):559-567. DOI: 10.2298/HEMIND120703097K

[32] Kostić MM, Hurt AP, Milenković DD, Velinov ND, Petrović MM, Bojić DV, et al. Effects of ultrasound on removal of ranitidine hydrochloride from water by activated carbon based on *Lagenaria siceraria*. Environmental Engineering Science. 2019;**36**(2):237-248. DOI: 10.1089/ees.2017.0539

[33] Fuertez-Córdoba JM, Acosta-Pavas JC, Ruiz-Colorado AA. Alkaline delignification of lignocellulosic biomass for the production of fermentable sugar syrups. DYNA. 2021;**88**(218):168-177. DOI: 10.7440/res64.2018.03

[34] Chaturvedi V, Verma P. An overview of key pretreatment processes employed for bioconversion of lignocellulosic biomass into biofuels and value added products. Biotech. 2013;**3**:415-431. DOI: 10.1007/s13205-013-0167-8

[35] Singh A, Sharma RK, Agrawal M, Marshall FM. Risk assessment of heavy metal toxicity through contaminated vegetables from waste water irrigated area of Varanasi, India. Tropical Ecology. 2010;**51**(2S):375-387

[36] Chang VS, Holtzapple MT. Fundamental factors affecting biomass enzymatic reactivity. Applied Biochemistry and Biotechnology. 2000;**84**:5-37. DOI: 10.1385/ABAB:84-86:1-9:5

[37] Marković-Nikolić DZ, ALj B, Savić SR, Petrović SM, Cvetković DJ, Cakić MD, et al. Synthesis and physicochemical characterization of anion exchanger based on green modified bottle gurd shell. Journal of Spectroscopy. 2018;**2018**:1-16. DOI: 10.1155/2018/1856109

[38] Peng BL, Dhar N, Liu HL, Tam KC. Chemistry and applications of cellulose nanocrystal and its derivatives: A nanotechnology perspective. Canadian Journal of Chemical Engineering. 2011;**89**(5):1191-1206. DOI: 10.1002/cjce.20554

[39] Eichhorn SJ. Cellulose nanowhiskers: Promising materials for advanced applications. Soft Matter. 2011;**7**(2): 303-315. DOI: 10.1039/C0SM00142B

[40] Kamel S. Nanotechnology and its applications in lignocellulosic composites, a mini review. Express

Polymer Letters. 2007;**1**(9):546-575. DOI: 10.3144/expresspolymlett.2007.78

[41] Bansiwal AK, Rayalu SS, Labhasetwar NK, Juwarkar AA, Devotta S. Surfactant-modified zeolite as a slow release fertilizer for phosphorus. Journal of Agricultural and Food Chemistry. 2006;**54**(13):4773-4779. DOI: 10.1021/jf060034b

[42] Namasivayam C, Sureshkumar MV. Removal of chromium(VI) from water and wastewater using surfactant modified coconut coir pith as a biosorbent. Bioresource Technology. 2008;**99**(7):2218-2225. DOI: 10.1016/j.biortech.2007.05.023

[43] Bingol A, Aslan A, Cakici A. Biosorption of chromate anions from aqueous solution by a cationic surfactant-modified lichen (*Cladonia rangiformis* L.). Journal of Hazardous Materials. 2009;**161**(2-3):747-752. DOI: 10.1016/j.jhazmat.2008.04.018

[44] Ibrahim S, Ang H-M, Wang S. Removal of emulsified food and mineral oils from wastewater using surfactant modified barley straw. Bioresource Technology. 2009;**100**(23):5744-5749. DOI: 10.1016/j.biortech.2009.06.070

[45] Chen H, Zhao J, Wu J, Dai G. Isotherm, thermodynamic, kinetics and adsorption mechanism studies of methyl orange by surfactant modified silkworm exuviae. Journal of Hazardous Materials. 2011;**192**(1):246-254. DOI: 10.1016/j.jhazmat.2011.05.014

[46] Syverud K, Xhanari K, Chinga-Carrasco G, Yu Y, Stenius P. Films made of cellulose nanofibrils: Surface modification by adsorption of a cationic surfactant and characterization by computer-assisted electron microscopy. Journal of Nanoparticle Research. 2011;**13**:773-782. DOI: 10.1007/s11051-010-0077-1

[47] Su Y, Zhao B, Xiao W, Han R. Adsorption behavior of light green anionic dye using cationic surfactant-modified wheat straw in batch and column mode. Environmental Science and Pollution Research. 2013;**20**(8):5558-5568. DOI: 10.1007/s11356-013-1571-7

[48] Foroughi-dahr M, Abolghasemi H, Esmaieli M, Nazari G, Rasem B. Experimental study on the adsorptive behavior of Congo red in cationic surfactant-modified tea waste. Process Safety and Environmental Protection. 2015;**95**:226-236. DOI: 10.1016/j.psep.2015.03.005

[49] Marković-Nikolić DZ, ALj B, Petković G, Ristić N, Cakić MD, Nikolić GS. The preparation and utilization of the cationic sorbent based on the surfactant modified bottle gourd shell. Advanced Technologies. 2017;**6**(2):38-50. DOI: 10.5937/savteh1702038M

[50] de Gennaro B, Catalanotti L, Bowman RS, Mercurio M. Anion exchange selectivity of surfactant modified clinoptilolite-rich tuff for environmental remediation. Journal of Colloid and Interface Science. 2014;**430**:178-183. DOI: 10.1016/j.jcis.2014.05.037

[51] Xu F, Yu J, Tesso T, Dowell F, Wang D. Qualitative and quantitative analysis of lignocellulosic biomass using infrared techniques: A mini-review. Applied Energy. 2013;**104**:801-809. DOI: 10.1016/j.apenergy.2012.12.019

[52] Kaur K, Phutela UG. Sodium carbonate pretreatment: An approach towards desilication of paddy straw and enhancement in biogas production. Paddy and Water Environment. 2016;**14**:113-121. DOI: 10.1007/s10333-015-0483-1

[53] Chintakunta R, Buaron N, Kahn N, Moriah A, Lifshiz R, Traitel T, et al.

Synthesis, characterization, and self-assembly with plasmid DNA of a quaternary ammonium derivative of pectic galactan and its fluorescent labeling for bioimaging applications. Carbohydrate Polymers. 2016;**150**:308-318. DOI: 10.1016/j.carbpol.2016.05.015

[54] McCormick CL, Dawsey TR. Preparation of cellulose derivatives via ring-opening reactions with cyclic reagents in lithium chloride/N,N-dimethylacetamide. Macromolecules. 1990;**23**(15):3606-3610. DOI: 10.1021/ma00217a011

[55] Xu X, Gao Y, Gao B, Tan X, Zhao Y, Yue Q, et al. Characteristics of diethylenetriamine-crosslinked cotton stalk/wheat stalk and their biosorption capacities for phosphate. Journal of Hazardous Materials. 2011;**192**(3):1690-1696. DOI: 10.1016/j.jhazmat.2011.07.009

[56] Wang Z, Bo N, Liu Y, Yang G, Liu Y, Zhao Y. Preparation of lignin-based anion exchangers and their utilization for nitrate removal. BioResources. 2013;**8**(3):3505-3517. DOI: 10.15376/biores.8.3.3505-3517

[57] Silverstein RM, Bressler GC, Morril TC. Spectrometric Identification of Organic Compounds. 5th ed. New Jersey: John Wiley & Sons; 2005

[58] Marković-Nikolić DZ, ALj B, Bojić DV, Cakić MD, Cvetković DJ, Nikolić GS. The biosorption potential of modified bottle gourd shell for phosphate: Equilibrium, kinetic and thermodynamic studies. Chemical Industry and Chemical Engineering Quarterly. 2018;**24**(4):319-332. DOI: 10.2298/CICEQ171019006M

[59] Marković-Nikolić DZ, Cakić MD, Petković G, Nikolić GS. Kinetics, thermodynamics and mechanisms of phosphate sorption onto bottle gourd biomass modified by (3-chloro-2-hydroxypropyl)trimethylammonium chloride. Progress in Reaction Kinetics and Mechanism. 2019;**44**(3):267-285. DOI: 10.1177/1468678319858149

[60] Nikolić GS, Marković-Nikolić DZ, Nikolić T, Stojadinović D, Andjelković T, Kostić M, et al. Nitrate removal by sorbent derived from waste lignocellulosic biomass of *Lagenaria vulgaris*: Kinetics, equilibrium and thermodynamics. International Journal of Environmental Research. 2021;**15**(1):215-230. DOI: 10.1007/s41742-021-00310-8

[61] Nikolić GS, Simonović N, Lj N, Durmišević M, Marković-Nikolić DZ, Ristić N, et al. An integrated OVAT-RSM design to gaps-filling in the study of phosphate sorption process onto cationic modified bottle gourd shell. Advanced Technologies. 2023;**12**(1):5-19. DOI: 10.5937/savteh2301005N

[62] Marković-Nikolić DZ, Bojić A, Bojić D, Cvetković D, Cakić M, Nikolić GS. Preconcentration and immobilization of phosphate from aqueous solutions in environmental ceanup by a new bio-based anion exchanger. Waste and Biomass Valorization. 2020;**11**(4):1373-1384. DOI: 10.1007/s12649-018-0401-z

Section 3

Sorption Processes for Environmental Protection and Pollution Control

Chapter 6

The Impact of Green Technology on Sorption Processes

Ijeoma J. Ani, Uduak G. Akpan, Ezeh E. Mbamalu and Chinedu T. Egbosiuba

Abstract

Sorption is the collective term used for both adsorption and absorption. Absorption involves the diffusion of molecules of a substance into a material due to the material's ability to dissolve the substance, while adsorption involves the attachment of molecules of a substance to a material surface. Sorption has been gaining significant attention in recent years as a promising separation technique because it is simple, efficient, and low-cost. Some new perspectives and applications of sorption are Sorption in Nanotechnology, water treatment, gas separation and Medicine and Biotechnology. Sorption is a promising alternative to traditional separation methods, which have significant economic and environmental drawbacks. Sorption can serve numerous applications across various sectors, including industry, medicine, and energy. This technique is easy to execute, energy-efficient, and economical, and has the potential to develop sustainable solutions through resource recycling, green technology and waste reduction. In this chapter, we discussed the impact of green technology on sorption processes, recent developments and the challenges that needs more research that can proffer solution and make sorption processes more reliable and attractive.

Keywords: adsorption, absorption, wastewater treatment, green technology, nanotechnology, sorption processes

1. Introduction

By IUPAC (1997/2019) definition, sorption is "The process by which a substance (sorbate) is sorbed (adsorbed or absorbed) on or in another substance (sorbent)." For adsorption, the International Adsorption Society (IAS, 2019) indicates that it is "The use of solids for removing substances from either gaseous or liquid solutions, and involves the preferential partitioning of substances from the gaseous or liquid phase onto the active part of a solid substrate" [1]. This process has been employed since eighteenth century [2], to proffer solutions to various environmental pollution challenges.

Pollutants from human activities, industries, pose a severe threat to humans and the eco-system. With the global increase in population, more pollution is encountered in various sectors which requires urgent and effective method for ablation of the pollutants. Industrial emissions are one of the primary air pollutants and waste fumes produced from several activities which can cause crisis to individual health and ecological damage [3]. Industrial air have volatile organic compounds (VOCs) and volatile inorganic compounds (VICs) which can cause odor and this problem is encountered in various industrial sectors such as petro-refineries, latex processing, bulk drug and pharmaceuticals, tanneries, waste treatment plants, poultry farms, and fish processing facilities [3]. These challenges can be successfully removed by sorption processes that are sustainable via green technology.

Absorption method can be used for the removal/control of gaseous pollutants. These pollutants could be acidic (hydrochloric acid, sulfuric acid, hydrogen sulfide), organic (Ethylene, Benzene, Ethanol, and many other volatile organic compounds (VOCs) or hazardous air pollutants (HAPs)). It involves the transfer of a gaseous pollutant from the air into a contacting liquid, such as water. The liquid must be able either to serve as a solvent for the pollutant or to capture it by means of a chemical reaction. On the other hand, adsorption, which is the separation process of guest molecules from the environment to the bulk or surface of the solid [4], involves the capture of the gaseous substance on an active part of an adsorbent.

Green technology is a term used to describe environmentally friendly processes and products that have a minimal impact on the environment. Adsorption is one of the most steadfast techniques employed as it is simple in operation, and cost-effective. Also, this technique provides a wider choice of material sources known as adsorbents which can be rationally designed to suit the application. To this degree, nature provides a greener choice of several adsorbents which are readily available and do not require high-end synthetic processes. These green sorbents are inclusive of natural materials, biomaterials, or biosorbents, carbonaceous materials from wastes, agricultural, and industrial by-products [5]. Another example of a green technology that uses absorption is absorption chillers. These chillers use water as the refrigerant, which makes them environmentally friendly and sustainable. Also, absorption method can be applied on CO_2 capture using green technology [6], likewise hydrogen production.

In recent research, nanotechnology has emerged as one of the most interesting areas due to its significant impact which is quite effective in sorption processes. Green technology has been applied for the production of nano-materials for various applications which is sustainable. It has been applied intensively in adsorption process with high efficiency, especially for the eradication of pollutants wastewater [7–9]. Preparation of nanomaterials using green technology is attractive due to their environmental and biocompatible features enabled by the use biological wastes, microorganisms, and different plant materials such as leaves, roots, and flowers [9]. In nanotechnology applications, the biomolecules serve as a reducing and stabilizing agent in the green production of nanomaterials [10]. For instance, magnetic nanohybrid adsorbent ($ZnCoFe_2O_4$@Chitosan (Ch)) was successfully synthesized for tetracycline adsorption [11]. $CuMn_2O_4$ nanostructures was synthesized using onions as the capping agent which helped to achieve the nanomaterials [12]. Thus, nanotechnology has paved way for a sustainable development in sorption processes with the application of green technology.

This chapter encompasses the recent trend in the sorption processes that have created opportunities for sustainable development.

2. Sorption processes

Sorption is a process that involves the adsorption and absorption of substances onto a solid or liquid surface. Adsorption refers to the attachment of molecules or ions from a gas or liquid onto the active part of a solid material [13]. The adsorption on the solid surface is that the molecules or atoms on the solid surface have residual surface energy due to unbalanced forces. When some substances undergo collision with solid surface, they are attracted by these unbalanced forces and stay on the solid surface [14]. The adsorbate molecules are held onto the surface by weak van der Waals forces or electrostatic interactions. Absorption, on the other hand, involves the uptake of a substance into the bulk of a solid or liquid material. This occurs when the absorbate molecules penetrate the surface of the absorbent and become incorporated within its structure. Adsorption can involve physical processes, such as diffusion into pores or interstitial spaces, or chemical reactions with the adsorbent material [15].

The sorption process can be used for various applications, including purification, separation, and storage of substances. For example, activated carbon is commonly used as an adsorbent material in water treatment processes to remove contaminants which have been deeply researched on. Sorption processes is equally applicable in gas separation, where specific adsorbents are used to selectively adsorb certain gases from a mixture. This is commonly seen in processes such as natural gas purification or carbon dioxide capture [16–18]. In drug delivery system, sorption technology is applicable [19]. It can be used to inject medicinal substance into the sorbent under the conditions of reversibility of the process and desorption of the medicinal substance into the organism. The sorbent is pre-saturated with the necessary medicinal substance and the system is used in the desorption mode [19]. Bio-materials can be used for drug delivery system [20] likewise other materials such as activated carbon, mineral sorbents, and polymers.

Materials used for sorption processes are cost effective, likewise the application of the material. Agricultural waste and a lot of biomaterials can be used for the production of adsorbent which is a sustainable. Also, natural clays or which are abundant in nature are good absorbent. The natural clay such as kaolinite can be modified to produce adsorbents such as zeolite which has proven so effective for sorption processes in the works that have been reported [21–23]. The natural clay can be used directly for adsorption processes or it can be activated to improve its activities as an adsorbent. This green technology adaptation in sorption process helps to cob challenges such as environmental hazards during chemical production of adsorbent, expensive chemicals for adsorbent production, instability of adsorbent.

3. Green technology for sustainable development

Green technology is an umbrella term that captures any technology that is created to be environmentally friendly from its production line all the way to its usage [24]. Also, it is known as sustainable technology or clean technology for the development and application of innovative solutions that minimize the negative impact on the environment while promoting sustainable development. It aims to address environmental challenges and create a more sustainable future by reducing resource consumption, minimizing pollution, and promoting renewable energy sources.

Green technology covers a wide range of sectors and industries, including energy, transportation, waste management, agriculture, and construction. Some examples of green technologies include but not limited to the following [25–27]:

1. Renewable energy: Technologies that harness energy from renewable sources such as solar, wind, hydro, and geothermal power, likewise biomass and biofuels. These technologies help reduce greenhouse gas emissions and dependence on fossil fuels.

2. Energy efficiency: Technologies that improve energy efficiency in buildings, appliances, and industrial processes. This includes energy-efficient lighting, insulation, smart grids, and energy management systems.

3. Waste management: Technologies that promote recycling, waste reduction, and waste-to-energy conversion. This includes recycling facilities, composting systems, and anaerobic digestion for organic waste.

4. Sustainable transportation: Technologies that reduce emissions and promote sustainable modes of transportation. This includes electric vehicles, hybrid vehicles, public transportation systems, and bike-sharing programs.

5. Water and air purification: Technologies that clean and purify water and air, reducing pollution and improving public health. This includes water filtration systems, air purifiers, and wastewater treatment plants.

6. Sustainable agriculture: Technologies that promote sustainable farming practices, reduce chemical inputs, and improve resource efficiency.

3.1 Green technology in sorption processes: it's new perspectives

One of the areas green technologies is applicable is the sorption process, which involves the removal of contaminants from wastewater, air, and gas streams. This process has gained much attention due to its potential to reduce pollution and safeguard human health. Green technology in sorption processes involves the use of sustainable, eco-friendly materials and methods for the removal of contaminants or for separation processes. This is the new perspective in our today sorption processes.

One of the green technologies that have shown great promise in sorption processes is the use of biochar. Biochar is a carbon-rich product that is derived from the pyrolysis of organic material, such as agricultural waste, manure, and wood chips. Biochar has a high surface area, which allows it to adsorb pollutants efficiently. Moreover, the use of biochar is environmentally friendly, as it reduces greenhouse gas emissions and mitigates climate change. A lot of reports have been made on the use of biochar for sorption processes [28–30]. However, biochar can post threats to the environment if not properly managed after use.

In addition to the use of sustainable sorbents, green technology in sorption also involves the use of renewable energy for the regeneration of sorbents. Lu et al. [31] successfully designed a Solar energy-triggered regenerative bionic fiber adsorbent for CO_2 capture through bionic electrostatic and chemical cross-linking assembly of cellulose nanofibers, thermosensitive polymer poly (N-isopropylacrylamide), nano-graphene oxide, and polyethyleneimine. The material was used for the regeneration

DOI: http://dx.doi.org/10.5772/intechopen.1005286

of adsorbent used for CO_2 capture. Regeneration is an essential process in sorption as it allows the sorbents to be reused. The use of renewable energy sources, such as solar energy for the regeneration process reduces the environmental impact of sorption processes and contributes to sustainable development.

Green technologies in sorption processes offer new perspectives for managing environmental concerns using sustainable and environmentally friendly methods. Researchers have developed new generation sorbents and sorption technologies that demonstrate selective adsorption of pollutants. These materials have enabled the effective removal of toxic and harmful pollutants from various streams while also producing less waste and reducing energy costs.

One notable development in green sorption technologies is the development of innovative adsorbents which are biodegradable and can be naturally derived. These adsorbents include biomass derived from algae, agricultural residues, and food waste, among others [32–34]. They have been shown to be effective in the removal of various pollutants and are highly efficient as compared to conventional sorbents. In absorption processes, green technology has been used for the production of membrane for CO_2 absorption [35]. Huang et al. [36] researched on the green preparation of hollow fiber membrane which can be used for gas separation using absorption process. **Figure 1** presents a sustainable utilization of adsorption materials for wastewater treatment.

Table 1 presents recent sorbents that are sustainable. It can be observed that green technology, which is a sustainable development, was applied for the development of the adsorbents. Most authors reported to have developed a biodegradable adsorbent with a renewable nature which offers sustainable solutions to address water pollution and contaminants. However, it could still take a long time before it can degrade. The method of synthesis of these adsorbents and the materials used for the synthesis are mostly cost-effective and eco-friendly. Also, stable adsorbents are being development which is beneficial for the economy. Phytochemicals from plants and other natural sources are used for the synthesis of nano-adsorbents which are in abundant and eco-friendly. It can be seen that most recent bio-adsorbent has high adsorption capacity which is commendable. The phytochemicals act as the capping or reducing agent which leads to the development of a nano-adsorbent.

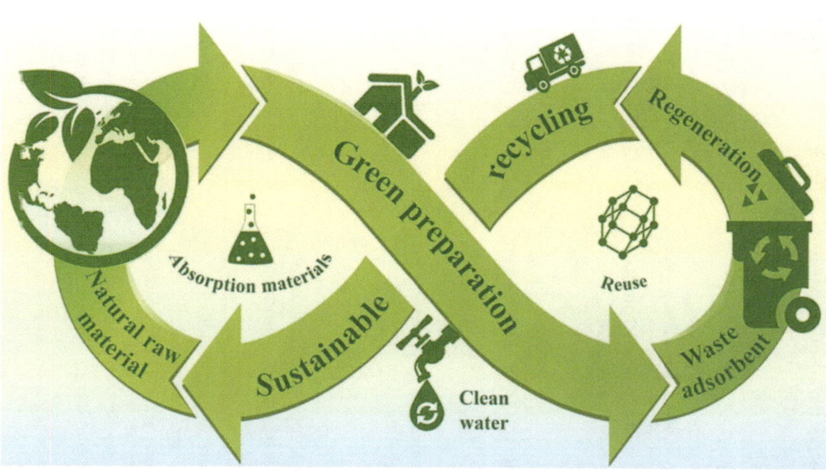

Figure 1.
Green technology; a sustainable utilization of adsorbent for wastewater treatment. (Source: Ref. [37]).

Adsorbent	Bio-source	Adsorption capacity (mg/g)	pollutant	Remark	Reference
Hyphaene thebaica derived-biochar (HTBC)	Egyptian doum palm	264.922	Methyl orange dye (MO)	Biodegradable adsorbent was successfully synthesized.	[38]
Curdlan/sodium carboxymethylcellulose composite (CURD/CMC)	Curdlan	385.85	Methylene blue (MB)	Biodegradable, observed synergistic effect between CURD and CMC with better performance.	[39]
Green nano-magnetic/biochar (BC) composite (nM-BC)	Tamarindus Indica fruit seed	BC; MB(3.055), malachite green (MG) (5.577). nM-BC; MB(3.326), MG (20.408)	MB and MG	Adsorbent easy to recover and it is reusable. Ability to adsorb both anionic and cationic dyes	[40]
Membrane polymeric adsorbent	Breadfruit, or Artocarpus altilis	176.25	Crystal violet (CV)	Biodegradable adsorbent. Reusable up to four cycles tested without any significant decrease in efficiency	[41]
Carbohydrate polymeric adsorbents	rice flour (RF) and graham flour (GF)	173.24(RF) and 151.27(GF)	MO	Biodegradable adsorbent, reusable and stable, The competing ions were not adversely affected in the dye adsorption as defined by the stable bonding mechanism	[42]
UV-photopolymerized cassava starch-based superabsorbent hydrogels	Cassava starch (CSt) and polyacrylic acid (PAA)	1044	MB	biodegradable, low-cost, and effective adsorbent. Excellent adsorption capacity. Also reusable with good biosafety	[43]
Sodium alginate-modified alkali-activated eggshell/Fe_3O_4 nanoparticles	Eggshell	332.57–256.62(CV) and 304.47–240.62(MB)	CV, MB	Easy to recover from the solution and it is reusable	[44]
All-biopolymer self-assembling hydrogel film	chitosan and carboxymethyl guar gum	155.51	Cu^{2+} ions	An eco-friendly bio-adsorbent, sustainable, and efficient. Also, it is reusable	[45]

DOI: http://dx.doi.org/10.5772/intechopen.1005286

Adsorbent	Bio-source	Adsorption capacity (mg/g)	pollutant	Remark	Reference
Fungi/MCM-41 bio-based adsorbent	Naturally available fungal wastes	94.75%, 60 min, 20 mg/L, 1.67 g/L	Reactive Red 120 (RR120)	A biocompatible, environmentally friendly, and inexpensive adsorbent was successfully synthesized which was effective. Also, the fungi is abundant in nature	[46]
Ceria (CeO_2) and areca nut shell biochar (CB) composite	Saccharum officinarum extract as a solvent/ capping agent	492.2	MB	Plant extract was use for the synthesis of the adsorbent which served as a capping and reducing agent	[47]
Modified magnetite	gaharu leaf extract	—	MO	The plant extract helped to modify the magnetite which improved the adsorption capacity. The adsorbent is reusable	[48]
Green functionalized magnetic graphene oxide	—	586	Silver	Nano-adsorbent was successfully synthesized which enhanced the performance of the adsorbent	[49]
Chitosan/functionalized fruit stones	Fruit stones	409.63	brilliant green (BG) dye	Remarkable adsorption capacity	[50]

Table 1.
Bio-adsorbents for sorption processes.

Another green technology that has gained attention in sorption processes is nanotechnology which happens to be among the fasted growing field for innovative development. Nanomaterials are tiny particles that have unique structural and chemical properties, which make them highly effective in the removal of contaminants. One example of such material is graphene oxide [51], which has shown great potential for the removal of organic pollutants from wastewater. The use of nanomaterials is also beneficial as it reduces the amount of sorbent needed, thus reducing waste generation.

3.2 Nanotechnology

Nanotechnology has been one of the most attractive research areas in most recent time due to its significant in sorption processes in respect to green technology. It is a field of science and technology that deals with the manipulation and control of matter at the nanoscale, which is the scale of atoms and molecules [52]. It involves the design, synthesis, characterization, and application of materials and devices with unique properties and functionalities at the nanoscale. In the context of green technology and sustainable development, nanotechnology offers several potential benefits such as energy efficiency, pollution prevention, waste reduction, environmental monitoring, and sustainable agriculture. In this chapter, we are more concern with nanotechnology on sorption processes.

3.3 Nanotechnology for sorption processes

Nanotechnology has the potential to revolutionize green technology by offering innovative solutions for various environmental challenges. Water purification: Nanotechnology offers advanced filtration and purification techniques for water treatment. Nanomaterials, such as graphene oxide and nanofibers, can effectively remove contaminants, including heavy metals, organic pollutants, and bacteria, from water. Nanotechnology-based membranes and adsorbents can provide high water permeability and selectivity, improving the efficiency of water purification processes. Recent research has shown magnificent effect of nanomaterial on the eradication of pollutants via adsorption processes. *Cola nitida* leave extract was used to synthesize green iron oxide nanaoparticles for the adsorption of dye pollutants [9]. *Psidium guajava* leaves extract was utilized to synthesize green nickel oxide nanoparticles (PG-NiONPs) [10]. Yahya et al. [53] investigated the characterization, preparation and application of organoclay nano-adsorbent for the removal of pollutants from wastewater. They concluded that the Organoclay could be considered a cheap and efficient adsorbent for the removal of most of the chemical pollutants from wastewater that could be of socioeconomic and environmental relevance. Also, a magnetic polyacrylonitrile-melamine nano-adsorbent (PAN-Mel@Fe_3O_4) has been developed as a polymer-based adsorbent for the adsorption of cadmium (Cd) and lead (Pb) from aqueous media [54]. Thus, nanotechnology has made remarkable impact in our society. The green technology has tackled the expensive method of nano-adsorbent production. However, during application, agglomeration of the nano-adsorbent may be observed which could reduce the effectiveness of the material. Also, it is possible to have some nano-adsorbent left in the treated effluent/water which is not favorable to the eco-system. **Figure 2** depicts nano-adsorbents and their applications.

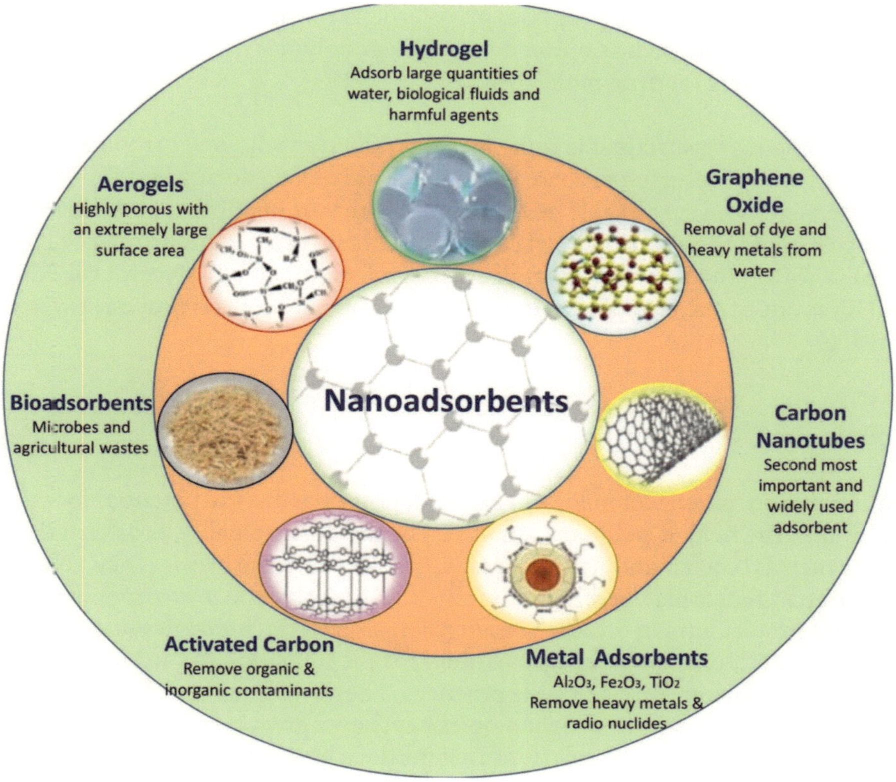

Figure 2.
Nanosorbents and their application (Source: Ref. [55]).

4. Sorption applications

Sorption processes, have various applications in different fields. Here are some examples:

1. Water treatment: Sorption processes are widely used in water treatment to remove contaminants from drinking water or wastewater. Green synthesized adsorbent has been used for the removal of all kinds of pollutants from water which have been reported [56].

2. Air purification: Sorption processes are employed in air purification systems to remove pollutants and improve indoor air quality. Activated carbon filters are commonly used to adsorb volatile organic compounds (VOCs), odors, and other harmful gases. Zeolites and molecular sieves can also be used to selectively adsorb specific gases, such as carbon dioxide or nitrogen oxides.

3. Gas separation: Sorption processes are utilized in gas separation applications, such as natural gas purification and carbon capture. Adsorbents, such as activated carbon or zeolites, can selectively adsorb certain gases, allowing for the separation and purification of gas mixtures. This is particularly important in industries like natural gas processing and biogas upgrading.

4. Drug delivery: Sorption processes play a crucial role in drug delivery systems. Porous materials, such as mesoporous silica or polymer nanoparticles, can be used as carriers for drug molecules.

Development of a sustainable route for H_2 production with low carbon emission is an urgent area of research to abate the high release of CO_2 during H_2 [57]. Recent findings has shown a sustainable development for the production of pure H_2 without emitting CO_2 into the atmosphere via Sorption-enhanced steam reforming of bio-ethanol (SESRE) and sorption enhanced steam reforming of bio-glycerol (SESRG) [57]. This is one of the new dimensions in recent time on sorption processes which is promising.

5. Conclusion

In conclusion, green technology in sorption processes has the potential to revolutionize the field of pollutant removal. The use of sustainable, eco-friendly sorbents such as biochar and nanomaterials, coupled with the use of renewable energy sources for sorbent regeneration, can significantly reduce environmental impact and protect human health. The adoption of green technologies in sorption processes, therefore, presents an opportunity for the transition toward more sustainable and environmentally responsible practices. The development of innovative sorbent materials and sorption technologies has shown great potential for sustainable and environmentally friendly solutions that produce less waste and reduce energy costs. The adoption of green sorption technologies, therefore, offers new possibilities for addressing environmental concerns. However, regeneration of adsorbent needs proper study to make it more cost-effective, environmentally friendly. Also, the biodegradability of adsorbents should be considered as an area of interest to ascertain the duration of the degradation process.

Author details

Ijeoma J. Ani[1*], Uduak G. Akpan[2], Ezeh E. Mbamalu[3] and Chinedu T. Egbosiuba[4,5]

1 Department of Chemical Engineering, Nasarawa State University, Keffi, Nigeria

2 Department of Chemical Engineering, Federal University of Technology, Minna, Nigeria

3 Department of Chemical Engineering, Federal University Otuoke, Bayelsa State, Nigeria

4 Department of Chemical Engineering, Chukwuemeka Odumegwu Ojukwu University, Uli Campus, Anambra, Nigeria

5 Artie McFerrin Department of Chemical Engineering, Texas A&M University, College Station, USA

*Address all correspondence to: gentle2015ijtessy@gmail.com

References

[1] Núñez-Delgado A, Álvarez-Rodríguez E, Fernández-Sanjurjo MJ, Fernández-Calviño D, Conde-Cid M, Arias-Estévez M. Introduction. In: Sorbents Materials for Controlling Environmental Pollution: Current State and Trends. Amsterdam, Netherlands: Elsevier; 2021. pp. 1-12. DOI: 10.1016/B978-0-12-820042-1.00018-3

[2] Rouqucrol J, Rodriguez-Reinoso F, Sing KSW, Unger KK, Robens E. Some Intriguing Items in the History of Adsorption. Amsterdam, Netherlands: Elsevier; 1994

[3] Gandu B, Gangagni Rao A, Cahan R. Air pollution control by using different types of techniques and sorbents. In: Sorbents Materials for Controlling Environmental Pollution: Current State and Trends. Amsterdam, Netherlands: Elsevier; 2021. pp. 575-594. DOI: 10.1016/B978-0-12-820042-1.00002-X

[4] Pourhakkak P, Taghizadeh A, Taghizadeh M, Ghaedi M, Haghdoust S. Fundamentals of adsorption technology. In: Interface Science and Technology. Amsterdam, Netherlands: Elsevier B.V; 2021. pp. 1-70. DOI: 10.1016/B978-0-12-818805-7.00001-1

[5] Rego RM, Kurkuri MD, Kigga M. Sustainable green approaches in sorption-based defluoridation: Recent progress. In: Green Technologies for the Defluoridation of Water. Amsterdam, Netherlands: Elsevier; 2021. pp. 141-174. DOI: 10.1016/B978-0-323-85768-0.00021-X

[6] Zhao Y, Dong Y, Guo Y, Huo F, Yan F, He H. Recent progress of green sorbents-based technologies for low concentration CO_2 capture. Chinese Journal of Chemical Engineering. 2021;**31**:113-125. DOI: 10.1016/j.cjche.2020.11.005

[7] Egbosiuba TC, Abdulkareem AS, Tijani JO, Ani JI, Krikstolaityte V, Srinivasan M, et al. Taguchi optimization design of diameter-controlled synthesis of multi walled carbon nanotubes for the adsorption of Pb(II) and Ni(II) from chemical industry wastewater. Chemosphere. 2021;**266**:128937. DOI: 10.1016/j.chemosphere.2020.128937

[8] Egbosiuba TC, Abdulkareem AS, Kovo AS, Afolabi EA, Tijani JO, Auta M, et al. Ultrasonic enhanced adsorption of methylene blue onto the optimized surface area of activated carbon: Adsorption isotherm, kinetics and thermodynamics. Chemical Engineering Research and Design. 2020;**153**:315-336. DOI: 10.1016/j.cherd.2019.10.016

[9] Alex Mbachu C, Kamoru Babayemi A, Chinedu Egbosiuba T, Ifeanyichukwu Ike J, Jacinta Ani I, Mustapha S. Green synthesis of iron oxide nanoparticles by Taguchi design of experiment method for effective adsorption of methylene blue and methyl orange from textile wastewater. Results in Engineering. 2023;**19**. DOI: 10.1016/j.rineng.2023.101198

[10] Chukwu Onu D, Kamoru Babayemi A, Chinedu Egbosiuba T, Onyinye Okafor B, Jacinta Ani I, Mustapha S, et al. Isotherm, kinetics, thermodynamics, recyclability and mechanism of ultrasonic assisted adsorption of methylene blue and lead (II) ions using green synthesized nickel oxide nanoparticles. Environmental Nanotechnology, Monitoring & Management. 2023;**20**. DOI: 10.1016/j.enmm.2023.100818

[11] Nasiri A, Golestani N, Rajabi S, Hashemi M. Facile and green synthesis of recyclable, environmentally friendly,

chemically stable, and cost-effective magnetic nanohybrid adsorbent for tetracycline adsorption. Heliyon. 2024;**10**. DOI: 10.1016/j.heliyon.2024.e24179

[12] Samadi Kazemi M, Sobhani A. $CuMn_2O_4$/chitosan micro/nanocomposite: Green synthesis, methylene blue removal, and study of kinetic adsorption, adsorption isotherm experiments, mechanism and adsorbent capacity. Arabian Journal of Chemistry. 2023;**16**. DOI: 10.1016/j.arabjc.2023.104754

[13] Abin-Bazaine A, Campos Trujillo A, Olmos-Marquez M. Adsorption isotherms: Enlightenment of the phenomenon of adsorption. In: Wastewater Treatment. London, UK: IntechOpen; 2022. DOI: 10.5772/intechopen.104260

[14] Hu H, Xu K. Physicochemical technologies for HRPs and risk control. In: High-Risk Pollutants in Wastewater. Amsterdam, Netherlands: Elsevier; 2019. pp. 169-207. DOI: 10.1016/B978-0-12-816448-8.00008-3

[15] Wang WX. Bioaccumulation and biomonitoring. In: Marine Ecotoxicology: Current Knowledge and Future Issues. Amsterdam, Netherlands: Elsevier; 2016. pp. 99-119. DOI: 10.1016/B978-0-12-803371-5.00004-7

[16] Pramathesh R, Arun S. Natural gas treatment using adsorptive separation. In: Natural Gas - Extraction to End Use. London, UK: InTechOpen; 2012. DOI: 10.5772/53269

[17] Gong F, Hou D, Xiang X, Zhong C, Qiang X. Mechanistic study of the effect of clay hydration swelling on CH_4/CO_2 competitive adsorption. Journal of CO_2 Utilization. 2024;**80**. DOI: 10.1016/j.jcou.2024.102666

[18] Sardar P, Bhattacharya G, Manna R, Raj S, Rahut S, Nath Samanta A. Excellent CO_2 adsorption performance of amine-impregnated highly porous ZIF-8 adsorbent: Experimental and isotherm modeling studies. Advanced Powder Technology. 2024;**35**. DOI: 10.1016/j.apt.2024.104344

[19] Bondarev AV, Zhilyakova ET. Use of sorption processes in the technology of drug delivery systems. Farmatsiya i Farmakologiya. 2019;7:4-12. DOI: 10.19163/2307-9266-2019-7-1-4-12

[20] Adeosun SO, Ilomuanya MO, Gbenebor OP, Dada MO, Odili CC. Biomaterials for drug delivery: Sources, classification, synthesis, processing, and applications. In: Adv Funct Mater. London, United Kingdom: IntechOpen; 2020. Available from: https://www.intechopen.com

[21] Cecilia JA, Vilarrasa-García E, Morales-Ospino R, Finocchio E, Busca G, Sapag K, et al. Kaolinite-based zeolites synthesis and their application in CO_2 capture processes. Fuel. 2022;**320**. DOI: 10.1016/j.fuel.2022.123953

[22] Ji B, Zhang W. Adsorption of cerium (III) by zeolites synthesized from kaolinite after rare earth elements (REEs) recovery. Chemosphere. 2022;**303**. DOI: 10.1016/j.chemosphere.2022.134941

[23] Chen Z, Li X, Liu H, Xu W, Yu J, Zang Y, et al. A novel and cost-effective synthesis of magnetic zeolite 4A using kaolinite and red mud for Sr(II) removal. Microporous and Mesoporous Materials. 2024;**370**. DOI: 10.1016/j.micromeso.2024.113069

[24] Zaid Qamar M, Ali W. Green technology and its implications worldwide. In: Advanced Functional Materials. 2021. Available from:

https://www.researchgate.net/publication/350443477

[25] Marra A, Antonelli P, Pozzi C. Emerging green-tech specializations and clusters – A network analysis on technological innovation at the metropolitan level. Renewable and Sustainable Energy Reviews. 2017;**67**:1037-1046. DOI: 10.1016/j.rser.2016.09.086

[26] Sabban A. Introductory chapter: Green computing technologies and industry in 2021. In: Green Computing Technologies and Computing Industry in 2021. London, United Kingdom: IntechOpen; 2021. Available from: https://www.intechopen.com

[27] Zhu L, Luo J, Dong Q, Zhao Y, Wang Y, Wang Y. Green technology innovation efficiency of energy-intensive industries in China from the perspective of shared resources: Dynamic change and improvement path. Technological Forecasting and Social Change. 2021;**170**. DOI: 10.1016/j.techfore.2021.120890

[28] Obonukut M, Alabi S, Jock A. Biochar from Cassava Waste: A Paradigm Shift from Waste to Wealth. London, United Kingdom: IntechOpen; 2022. Available from: https://www.intechopen.com

[29] Valencia A, Muñiz-Valencia R, Ceballos-Magaña SG, Rojas-Mayorga CK, Bonilla-Petriciolet A, González J, et al. Cyclohexane and benzene separation by fixed-bed adsorption on activated carbons prepared from coconut shell. Environmental Technology and Innovation. 2022;**25**. DOI: 10.1016/j.eti.2021.102076

[30] Ma W, Han R, Zhang W, Zhang H, Chen L, Zhu L. Magnetic biochar enhanced copper immobilization in agricultural lands: Insights from adsorption precipitation and redox. Journal of Environmental Management. 2024;**352**. DOI: 10.1016/j.jenvman.2024.120058

[31] Lu W, Shi X, Zhou H, Luo W, Wang L, He H. Tailoring and properties of a novel solar energy-triggered regenerative bionic fiber adsorbent for CO_2 capture. Chemical Engineering Journal. 2022;**449**. DOI: 10.1016/j.cej.2022.137885

[32] El-Sheekh MM, El-Chaghaby GA, Rashad S. Bioenergy production from algae: Biomass sources and applications. In: Green Approach to Alternative Fuel for a Sustainable Future. Amsterdam, Netherlands: Elsevier; 2023. pp. 59-69. DOI: 10.1016/B978-0-12-824318-3.00032-1

[33] Mpatani FM, Han R, Aryee AA, Kani AN, Li Z, Qu L. Adsorption performance of modified agricultural waste materials for removal of emerging micro-contaminant bisphenol a: A comprehensive review. Science of the Total Environment. 2021;**780**. DOI: 10.1016/j.scitotenv.2021.146629

[34] Mumtaz H, Farhan M, Amjad M, Riaz F, Kazim AH, Sultan M, et al. Biomass waste utilization for adsorbent preparation in CO_2 capture and sustainable environment applications. Sustainable Energy Technologies and Assessments. 2021;**46**. DOI: 10.1016/j.seta.2021.101288

[35] Figueroa JR, Cuenca HE. Membrane gas absorption processes: Applications, design and perspectives. In: Osmotically Driven Membrane Processes - Approach, Development and Current Status. London, UK: InTechOpen; 2018. DOI: 10.5772/intechopen.72306

[36] Huang Y, Xiao C, Huang Q, Liu H, Zhao J. Progress on polymeric hollow

fiber membrane preparation technique from the perspective of green and sustainable development. Chemical Engineering Journal. 2021;**403**. DOI: 10.1016/j.cej.2020.126295

[37] Wang W, Wang A. Perspectives on green fabrication and sustainable utilization of adsorption materials for wastewater treatment. Chemical Engineering Research and Design. 2022;**187**:541-548. DOI: 10.1016/j.cherd.2022.09.006

[38] Tcheka C, Conradie MM, Assinale VA, Conradie J. Mesoporous biochar derived from Egyptian doum palm (*Hyphaene thebaica*) shells as low-cost and biodegradable adsorbent for the removal of methyl orange dye: Characterization, kinetic and adsorption mechanism. Chemical Physics Impact. 2024;**8**. DOI: 10.1016/j.chphi.2023.100446

[39] Cui C, Li D, Jun Wang L, Wang Y. Curdlan/sodium carboxymethylcellulose composite adsorbents: A biodegradable solution for organic dye removal from water. Carbohydrate Polymers. 2024;**328**. DOI: 10.1016/j.carbpol.2023.121737

[40] Bayram O, Moral E, Köksal E, Göde F, Pehlivan E. Removal of methyl blue and malachite green from water using biodegradable magnetic *Tamarindus Indica* fruit seed biochar: Characterization, equilibrium study, modelling and thermodynamics. Sustainable Chemistry for the Environment. 2023;**3**:100023. DOI: 10.1016/j.scenv.2023.100023

[41] Radoor S, Jayakumar A, Karayil J, Kim JT, Siengchin S. Biodegradable polymeric green adsorbent for the highly efficient removal of crystal violet dye from aqueous solution. Chemical Engineering Research and Design. 2023;**199**:473-485. DOI: 10.1016/j.cherd.2023.09.048

[42] Munjur HM, Hasan MN, Awual MR, Islam MM, Shenashen MA, Iqbal J. Biodegradable natural carbohydrate polymeric sustainable adsorbents for efficient toxic dye removal from wastewater. Journal of Molecular Liquids. 2020;**319**. DOI: 10.1016/j.molliq.2020.114356

[43] Srikaew M, Jumpapaeng P, Suwanakood P, Kaiyasuan C, Promarak V, Saengsuwan S. Rapid synthesis and optimization of UV-photopolymerized cassava starch-based superabsorbent hydrogels as a biodegradable, low-cost, and effective adsorbent for MB removal. Journal of Industrial and Engineering Chemistry. 2023;**118**:53-69. DOI: 10.1016/j.jiec.2022.10.045

[44] Elshimy AS, Mobarak M, Ajarem JS, Maodaa SN, Bonilla-Petriciolet A, Li Z, et al. Sodium alginate-modified alkali-activated eggshell/Fe_3O_4 nanoparticles: A magnetic bio-based spherical adsorbent for cationic dyes adsorption. International Journal of Biological Macromolecules. 2024;**256**. DOI: 10.1016/j.ijbiomac.2023.128528

[45] Rahmatpour A, Alijani N. An all-biopolymer self-assembling hydrogel film consisting of chitosan and carboxymethyl guar gum: A novel bio-based composite adsorbent for Cu2+ adsorption from aqueous solution. International Journal of Biological Macromolecules. 2023;**242**. DOI: 10.1016/j.ijbiomac.2023.124878

[46] Şimşek V, Yarbay RZ, Marttin V, Gül ÜD. Treatment of textile dye via economic fungi/MCM-41 bio-based adsorbent: Application of neural network approach. Journal of Cleaner Production. 2023;**421**. DOI: 10.1016/j.jclepro.2023.138448

[47] Gupta SV, Kulkarni VV, Ahmaruzzaman M. Fabrication of a bio-adsorbent material by grafting CeO_2 quantum dots (QDts) over areca nut shell biochar using *Saccharum officinarum* extract as a solvent/capping agent for adsorption of methylene blue dye: Synthesis, material analyses, adsorption kinetics and isotherms studies. Colloids and Surfaces A: Physicochemical and Engineering Aspects. 2024;**680**. DOI: 10.1016/j.colsurfa.2023.132611

[48] Sulistyaningsih T, Sari DA, Widiarti N, Astuti W, Wulandari R, Harjunowibowo D. Green synthesis of gaharu leaf extract-modified magnetite as an adsorbent for methyl orange textile dyes. Waste Management Bulletin. 2024;**2**:327-339. DOI: 10.1016/j.wmb.2024.02.007

[49] Zounia M, Yazdi MRS, Hakimi M, Zare H, Amiri A. Silver adsorption from gold leaching cyanide wastewater by a green functionalized magnetic graphene oxide nano-adsorbent. Hydrometallurgy. 2024;**225**. DOI: 10.1016/j.hydromet.2024.106265

[50] Khan MKA, Abdulhameed AS, Alshahrani H, Algburi S. Chitosan/functionalized fruit stones as a highly efficient adsorbent biomaterial for adsorption of brilliant green dye: Comprehensive characterization and statistical optimization. International Journal of Biological Macromolecules. 2024;**263**. DOI: 10.1016/j.ijbiomac.2024.130465

[51] Kumar B. Graphene-and Graphene Oxide-Bounded Metal Nanocomposite for Remediation of Organic Pollutants. London, UK: IntechOpen; 2020. Available from: https://www.intechopen.com

[52] Ani IJ, Egbosiuba TC. Metal oxides and their nanocomposites in wastewater treatment. In: Adsorption through Advanced Nanoscale Materials: Applications in Environmental Remediation. Amsterdam, Netherlands: Elsevier; 2023. pp. 185-203. DOI: 10.1016/B978-0-443-18456-7.00009-2

[53] Yahya K, Hamdi W, Hamdi N. Organoclay Nano-Adsorbent: Preparation, Characterization and Applications. London, United Kingdom: IntechOpen; 2022. Available from: https://www.intechopen.com

[54] Dogari H, Salehi MM, Hassanzadeh-Afruzi F, Saeidirad MM. Maleki, magnetic polyacrylonitrile-melamine nanoadsorbent (PAN-Mel@ Fe_3O_4) for effective adsorption of Cd (II) and Pb (II) from aquatic area. Materials Science & Engineering B. 2023;**298**:1-13

[55] Jain A, Kumari S, Agarwal S, Khan S. Water purification via novel nano-adsorbents and their regeneration strategies. Process Safety and Environmental Protection. 2021;**152**:441-454. DOI: 10.1016/j.psep.2021.06.031

[56] Hussain OA, Hathout AS, Abdel-Mobdy YE, Rashed MM, Abdel Rahim EA, Fouzy ASM. Preparation and characterization of activated carbon from agricultural wastes and their ability to remove chlorpyrifos from water. Toxicology Reports. 2023;**10**:146-154. DOI: 10.1016/j.toxrep.2023.01.011

[57] Xu Y, Wu M, Yang X, Sun S, Li Q, Zhang Y, et al. Recent advances and prospects in high purity H2 production from sorption enhanced reforming of bio-ethanol and bio-glycerol as carbon negative processes: A review. Carbon Capture Science & Technology. 2023;**8**:1-20

Chapter 7

Carbon Capturing Materials: A Review of Their Recent Modification Approaches and Implementation for Industrial Applications

Himangi Neve, Deven Kadam, Chinmay Sadadekar, Vinayak Wadgaonkar, Siraj Bhatkar and Prasad Wadekar

Abstract

An exponential rise in utilizing fossil fuels for the past 100 years has escalated the atmospheric concentration of CO_2 level, leading to the greenhouse effect. This problem has ignited many scientific minds to develop materials, methodologies, and technologies to reduce CO_2 emissions into the atmosphere. The know-how of properties affecting CO_2 capture mechanisms for zeolites, silica, graphene, activated carbon, MOFs, and layered double hydroxides (LDHs) are discussed. Further, we perform a comparative analysis of these conventional CO_2 adsorbing materials and their limitations with their modified counterparts. We explored the emerging hybrid composite materials, and their advancement over the modified materials is noteworthy and is highlighted in this review. Lastly, each of these modified CO_2 capture materials is discussed in relation to their relevance to industrial applications.

Keywords: GHG emissions, CO_2 capture, LDH, physical adsorption, coprecipitation

1. Introduction

Increasing evidence suggests that global climate change is caused by human-caused CO_2 emissions into the atmosphere. Capturing and storing combustion CO_2 reduces CO_2 emissions. Carbon capture is essential to the sustainability of carbon-intensive industries like the production of steel and cement. Carbon capture and utilization (CCU) is being considered as a technical solution for addressing problems related to greenhouse gas emissions and the rising global energy needs. CCS, or carbon capture and storage, is the process of isolating carbon dioxide (CO_2) generated by industrial and transportation activities and transporting it to a designated storage site [1, 2]. Five cutting-edge technologies—adsorption, membranes, cryogenics, microbial or algae—are frequently used to separate and capture CO_2.

The adsorption method has drawn a lot of attention because it is inexpensive, highly efficient, reversible, simple to use, and readily available adsorbent. The effectiveness and economics of adsorption processes such as pressure/temperature swing adsorption (PSA/TSA) are significantly influenced by the properties of the adsorbents, in addition to process design and operation parameters [3]. Adsorption has several benefits, including reduced energy consumption, economic regeneration, and a high loading capacity under the right circumstances. Additional benefits include a high adsorption rate, easy system maintenance, simple operation, good mechanical and chemical stability, and resistance to flue gas impurities [4].

Various solid adsorbents such as activated carbon, zeolites, and metal-organic frameworks have been used for capturing CO_2. These adsorbents often exhibit low adsorption effectiveness at elevated temperatures, usually over 100°C, for gases. Therefore, it is crucial to create advanced adsorbents that can effectively operate at elevated temperatures to capture CO_2. In this review, we summarize the basic working mechanism of these materials and their important properties, which affect the CO_2 capture process. Also, we highlight the recent advances carried out for these materials to enhance the CO_2-capturing process and their implementation for various industrial applications.

2. CO_2 adsorbents

Each of the adsorbents comes with unique properties that make them suitable for different applications in carbon capture (**Figure 1**). It is important to differentiate between amorphous and crystalline solids when discussing sorbents since their different physical and chemical properties have a substantial influence on CO_2 absorption ability.

2.1 Zeolites

Zeolites are naturally formed microporous crystalline silicate framework materials that can be created in a lab setting. Their homogeneous pore diameters range from

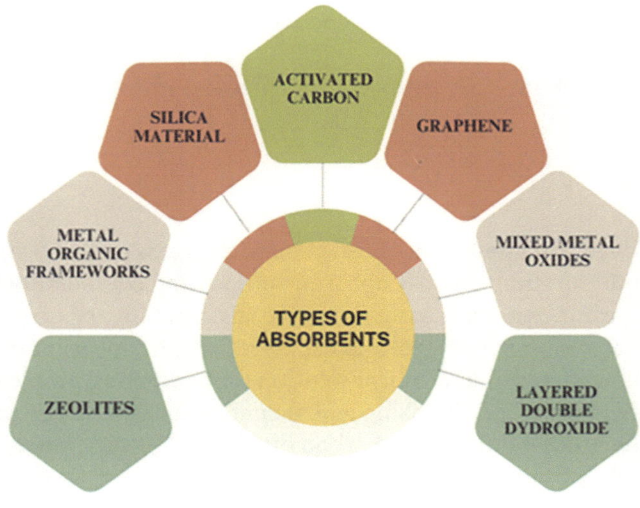

Figure 1.
Types of adsorbents.

0.5 to 1.2 nm, creating networks of interconnected channels or cages that adsorb gas molecules. Natural zeolite is a porous crystalline inorganic polymer that contains SiO_4 and AlO_4 tetrahedra (alumina silicate) with a three-dimensional structure [5]. This structure is permeated by water molecules and ions, allowing for significant mobility. Environmental applications, like catalysis and purification of water. This is because of its negative charge which results from replacing the Al^{3+} by Si^{4+} ions. The exchangeable cations neutralize the negative charge. It also has commendable adsorption power, which is credited to its high surface area and porous nature. CO_2 capture depends both on its cationic nature and the concentration of Al atoms present in the zeolite framework [6]. The CO_2 adsorption capacity of zeolites is significantly influenced by temperature and shows a positive correlation with their pore size. Zeolites' CO_2 adsorption ability diminishes notably when the temperature surpasses 100°C. An experiment was conducted to capture CO_2 at a temperature of 25°C using 0.5 g of adsorbent and a 15 vol% CO_2 mixture. Alkali metal-modified zeolite beta (Si/Al molar ratio 7.4) demonstrated that the adsorption capacity increased as the following ions were introduced: $Mg^{2+} < Ca^{2+} < Na^{+} < K^{+}$ ion [7]. Clinoptilolite, a natural zeolite, and linde type A zeolite (LTA), a synthetic zeolite, may both adsorb CO_2 via Van der Waals forces and modify their chemical composition (i.e., changing the type of the cations present in the framework) may increase their CO_2 adsorption capability. At 1000 mbar and 298 K, the CO_2 capture capability of LTA samples decreases significantly at low pressures (100 mbar), where the CO_2 adsorption of Ca-LTA and Na-LTA decreases by approximately 90% and 87%, respectively, from 298 to 423 K. The /kg CO_2 capturing cost of Clino is four times lesser compared to standard zeolite 13 X. Natural Clinoptilolite, which costs an average of $3 per kilogram of captured CO_2, may be an intriguing candidate for CO_2 capture up to 423 K [8].

2.2 Metal-organic frameworks (MOF)

MOFs are crystalline porous compounds composed of multidentate organic molecules that stabilize a three-dimensional network of metal ions. MOFs consist of metal-containing nodes that are typically assembled via strong coordination bonds and connected by organic ligand bridges. Due to the intertwined nature of the molecules, MOF adsorbents for CO_2 sequestration require spaces in the center of their structure. This sort of shape has a huge frame because of the connections between the pores and organic molecules and knots of metal ions. This type of shape has a very large frame extent. MOFs are easier to design, synthesize, and tune [9]. Adsorption temperature must be low enough to obtain a high CO_2 adsorption. The synthesis method of MOFs is expensive, which makes them economically impractical. The performance of MOFs has been notably improved by way of the MOFs mixed with the extraordinary functional materials and hence the new functionality changed into evolved in practical applications. MOFs exhibit a substantial reduction in adsorption capacity when exposed to gas mixtures [10]. At 2 bar and room temperature, ionic liquid (IL)/MOF membranes exhibit greater selectivity for CO_2/N_2 than the pristine ZIF-69 membrane. The H_2/CO_2 ideal separation factor of a ZIF-9 membrane coated with IL-functionalized carbon nanotubes was 40 at 1 bar, 298 K, and it retained its gas permeability and selectivity for 500 hours, indicating a high degree of stability. The solid-liquid transition and subsequent quenching of the liquid MOF result in the formation of MOF glasses. By utilizing an in situ solvothermal synthesis method to melt-quench a polycrystalline ZIF-62 membrane on a porous ceramic alumina support, a novel MOF glass membrane was fabricated. This membrane is devoid of grain boundaries and exhibits exceptionally high separation factors for CO_2/CH_4 and CO_2/N_2 at 1 bar and 298 K [11].

2.3 Silica

The potential uses of ordered mesoporous silica materials as solid sorbents for CO_2 collection have expanded across a wide range of scientific and technological fields due to their special qualities. They have a number of benefits, such as low manufacturing cost, fast kinetics, tunable pore size, outstanding regeneration stability characterized by substantial pore volumes and surface areas, low energy demands for regeneration, and favorable selectivity toward N_2 and CH_4. Additionally, robust thermal and mechanical stability is present. For improved CO_2 adsorption, mesoporous silica materials must be functionalized with different amine functional groups because their hydrothermal stability is significantly compromised when exposed to steam or scalding water. The advantages of aminosilane-based materials are their large CO_2 adsorption capacity and multicycle stability, which can be recycled. Mesoporous silica exhibits a remarkable propensity for adsorption when subjected to ambient temperature and arid, high-pressure conditions (approximately 45 bar). The large pore volume, high surface area, and thermal stability of mesoporous silica materials enable the immobilization of amine groups, thereby increasing their capacity for adsorbing CO_2 at low temperatures. However, upon loading the amines, the volume of the pore significantly reduces due to pore blocking. This obstruction hinders the diffusion of CO_2 through the material, consequently limiting the interaction between CO_2 and the interior amine groups. Consequently, this impacts the stability of the material throughout the adsorption-desorption cycle [12, 13]. The process of incorporating amines into mesoporous silica has been utilized to produce adsorbents that maintain their capacity even in the presence of moisture. In fact, moisture enhances their capacity rather than diminishing it. The pore surface of MCM-48 and SBA-15 silicas was modified using novel derivatives of (3-aminopropyl)triethoxysilane (APTES) with customized structures. These derivatives contained either an azomethine group that was linked to a furanyl ring or an aminophosphonate fragment. Each of the altered samples exhibited a 1.5- to 2-fold augmentation in the ability to adsorb CO_2 when compared to the original materials. The complete release of CO_2 occurred at a temperature of 80°C for the 3-[2-(2-aminoethylamino) ethylamino] propyl-trimethoxysilane (DAPTES)-modified mesoporous silicas, specifically SBA-15, SBA-16, and KIT-6 treated with DAPTES. The SBA-15 silica exhibited the highest CO_2 adsorption capability, measuring 3.9 mmol/g. The inclusion of 1 volume percent of water vapor resulted in an increase in the adsorption capabilities of the modified materials. The silica materials obtained exhibited consistent performance during five cycles of CO_2 adsorption and desorption. This suggests that the changed pore surface of the material has a significant interaction, likely involving the chemisorption of CO_2. This contact results in the creation of bicarbonates, which are then attracted to the positively charged ammonium groups by electrostatic forces [14].

2.4 Activated carbon (AC)

AC can be synthesized from any organic material that is carbon-rich and low in organic volatile strength. Biomass sources are naturally occurring, easily accessible, renewable, and economically viable. Activated carbons (ACs) have a large surface area and adsorption capability. To maintain the physicochemical characteristics,

DOI: http://dx.doi.org/10.5772/intechopen.1005188

originators of activated carbon should have high carbon and low ash concentrations, low volatile substances, and low propensity of degradation. Furthermore, the meso-porous/microporous structure of activated carbon allows for quick adsorption and lower energy consumption during the regeneration process [15]. Since chemically activated carbons have a larger surface area and more O functionalities groups on their adsorbent surface, they are able to absorb more CO_2 than physically activated carbons. By breaking down methane, scientists were able to alter a commercial AC surface, creating tiny micropores that can absorb CO_2 at concentrations as high as 1.7275 mmol/g at 273 K and 1 bar. Therefore, with heat of adsorption of 40.37 kJ/mol, the altered surface structure increased CO_2 selectivity over N_2 to a maximum of 22.84 [16]. Adsorption occurs in a CO_2/CH_4 mixture because of CO_2's strong quadrupole moment and CH_4's nonpolarity [17].

Under both pre- and post-combustion conditions, another set of researchers developed a very porous carbon adsorbent that can remove CO_2 from three distinct gas mixtures: CO_2/CH_4, CO_2/H_2, and CO_2/N_2. With a CO_2 absorption of up to 22.1 mmol/g at 25 pressure and remarkable selectivity against other gases, the created adsorbent demonstrated outstanding adsorption ability. The high selectivity and substantial CO_2 uptake are due to the carbon adsorbent's very high surface area (3010 m^2/g), total pore volume (1.053 cm^3/g), and the substantial O content on its surface, which offers CO_2 molecules attractive functional groups [18].

Since metallic oxides are electron donors, adding CuO to a carbon structure might increase the absorption capacity for acid gases like CO_2. Metal impregnation as a CO_2 adsorption method is rare, in contrast to N-doping technology [19].

A new CO_2 adsorbent with a suitable pore structure and an extremely high surface area is the graphene-AC composite. As an example, a composite material known as AC-PVA-GNP is created by hydrothermally combining AC and graphene nanoplatelets with polyvinyl alcohol. In contrast to pure AC, this composite had a greater CO_2 adsorption capacity of 1.347 cm^3/g and an extremely large surface area of 2730 m^2/g [20].

2.5 Graphene

Characteristics such as durability, large surface area, fast adsorption kinetics, and in-cycle operations support carbon-based adsorbents as potential CO_2-seizing adsorbents. The CO_2 adsorption capacity of ethylenediamine (EDA)-intercalated graphene oxide for CO_2/N_2 mixed gases was 4.65 wt.% at 303 K and 1 bar [13]. Nanoplates of graphene exfoliated thermally have the potential to function as innovative and exceptionally efficient CO_2 sorbents. The graphene nanoplates that were generated demonstrated a capture capacity of 248 wt% at 298 K and 30 bar [15].

2.6 Layered double hydroxides (LDH)

Layers of positively charged metal hydroxides resembling brucite comprise LDHs, with charge-balancing anions and water situated in the interlayer. It is possible to precisely regulate the chemical composition of the strata and the anions that move between them. Natural processes or precipitation constantly introduce organic or inorganic anions between the hydroxide layers [21].

The morphology of LDHs is mainly categorized into two types: powdery LDHs and spherical LDHs. The powdery LDHs commonly show brucite-like structure, which are mainly manufactured by the coprecipitation method. The spherical LDHs are commonly prepared by template-directed synthesis. Spherical LDHs have a variety of applications as compared to powdery LDHs like catalysts, sorbents, lithium-ion batteries, electrodes, and carriers for cellular drug and gene delivery [22].

LDHs are effective adsorbents for organic contaminants due to their low cost, positively charged layers with anion exchange capability, colloidal and thermal behavior, and large surface area. Homogeneous support modification can also improve the performance, thermal stability, morphology, and endurance of LDH by affording the opportunity to support the homogeneous reaction over a hybridized LDH [23].

Favorable characteristics of LDH for carbon capture:

1. High selectivity for CO_2.

2. High CO_2 working capacity between the conditions of adsorption and regeneration.

3. Mild conditions for regeneration.

4. High stability, mechanical strength, and resistance against impurities.

5. Moisture.

6. Fast adsorption and desorption kinetics.

7. Low cost.

LDHs, which have been developed as high-temperature (200–500°C) CO_2 adsorbents, have garnered considerable interest owing to their adjustable semiconductor properties and high CO_2 adsorption capacities in the interlayer space. Over the past decade, there has been considerable interest in these characteristics of LDHs, and they are now regarded as an extraordinary candidate for photoreduction and CO_2 capture. The approaches are primarily categorized as (1) solid phase synthesis and (2) liquid phase synthesis, depending on which phase is involved. The mechanochemical process used in the solid-state synthesis approach uses both wet and dry milling techniques. The Mulliken population of clusters, cell angles, bond lengths, and cation electronic structures all contribute significantly to the stability of LDHs. Capabilities can be increased by incorporating amino acids or amino silanes into stable layers and reacting with hydroxyl groups on laminates to generate an amine loading [24]. Research on CO_2 adsorption for amine-grafted LDH is not as extensive as it is for the extensively investigated LDH materials.

Water (H_2O) is found in both precombustion operations for H_2 generation and in the exhaust streams from coal-fired power stations in the post-combustion stage. In these processes, neither pressure nor temperature are significant concerns. Water molecules undergo a chemical reaction with metal oxides, resulting in the production of hydroxides. These hydroxides then interact with CO_2 to form bicarbonates [25].

When $Ti_1Li_3Al_2$-LDHs are subjected to various calcination temperatures, the plate structure of the LDHs is altered to varying extents. Moreover, once the configuration of LDHs starts to deform, the precise surface area of CO_2 progressively augments. The augmentation of the surface area, together with the reduction in the semiconductor band gap of LDHs, facilitates the transformation of CO_2 into methane [26].

Another study demonstrates a novel method in which a monolayer of terminal amino groups of aminopropylsilane (APS) is attached to LDHs (APS/LDHs) for the

Adsorbents	Advantages	Disadvantages
Zeolites	• High adsorption capacity • Regenerability • Stability • Wide availability	• Low CO_2 uptake at low pressure • Moisture sensitivity • Poor adsorption capacity at moderate temperature • Low selectivity toward CO_2
MOF	• High surface area • Tailorability • Large pore volumes	• Hydrothermal stability & durability • Poor economic efficiency • Material robustness • Scale-up challenges • Limited commercial availability • Complicated synthetic process
Silica materials	• Abundance and low cost • Chemical stability • Versatility	• Low CO_2 adsorption capacity • Limited selectivity • Kinetic limitations • Energy intensive regeneration
Activated carbon	• High adsorption capacity • Cost-effectiveness • Ease of regeneration • Proven technology	• Low selectivity • Vulnerability to contaminants • Limited reusability
Graphene	• Chemical inertness • Strong adsorption capacity • Selective adsorption • Unique structure with a high surface area	• Aggregation issues • Regeneration challenges • Scalability and cost
MMO	• Relatively fast adsorption/desorption rates • Good mechanical strength in high-pressures	• Weak adsorption capacity, although easily enhanced with potassium impregnation

Table 1.
Comparison of various adsorbents.

purpose of capturing CO_2. The increased space between layers, achieved by using dodecyl sulfonate as a pillar, is advantageous for the introduction of APS. APS is chemically bonded to the LDH laminates through covalent contact. The capacity of CO_2 adsorption reaches a stable level of around 90 mg g^{-1} over the course of five cycles of adsorption and desorption. This indicates a significant potential for application in temperature swing adsorption systems (**Table 1**) [27].

3. LDH synthesis methods

Generally, the LDH synthesis procedures are classified into one-step and multi-step methods. The one-step method includes coprecipitation and hydrothermal processes, which always achieve LDHs by direct preparation process. However, the multi-step method includes urea hydrolysis, the sol-gel method in which the exchange of ions and structural reconstruction takes place.

3.1 Urea hydrolysis

To improve the crystallinity of LDHs, this technique is frequently applied. Large particle sizes of well-crystallized LDHs are necessary to comprehend the basic LDH properties, as opposed to the nanoparticles or nanosheets of LDH in the dispersed medium. Large-sized, well-crystallized LDHs can be obtained through the homogeneous precipitation of LDHs utilizing the urea hydrolysis technique. The hydrothermal approach produces well-crystallized layered double hydroxides with a consistent composition, which gives it an edge over other procedures. Urea is used as the precipitating agent in urea hydrolysis. While urea can be substituted with NaOH, urea is preferred since it acts slowly and results in less supersaturation during precipitation. Urea is an exceedingly weak Bronsted base that exhibits high solubility in water and, when hydrolyzed in aqueous solutions under controlled conditions, can generate ammonium cyanate (or its ionic form: NH^{4+}, NCO^{-}) [28]. The layer-by-layer assembly of multilayer composite films, the creation of highly orientated films, and investigations into the mechanism of delamination have all utilized these crystalline LDHs. The sourcing and production of urea can contribute to greenhouse gas emissions and other environmental concerns. Utilizing urea, which is made from carbon dioxide and ammonia, could have an impact on the environment and product purity.

3.2 Sol-gel method

Several advantages distinguish this method from other conventional procedures: it is a straightforward method for obtaining nanoscale particles, which is essential for the majority of novel LDH applications, including the successful casting of LDH thin films. Purity-enhancing products are attainable. Mild conditions prevail for the duration of the procedure, facilitating the eventual intercalation of fragile organic molecules and biological entities. Polarity, reactivity, and quantity of the intended solvent (typically alcohol) are critical factors to take into account when devising an appropriate sol-gel methodology to enable the synthesis of pure LDHs featuring modifiable chemical compositions [29]. The sol-gel method may include the use of corrosive chemicals, high temperatures, and organic solvents, some of which may be hazardous or environmentally unfriendly.

3.3 Aqueous miscible organic solvent treatment (AMOST)

AMOST method uses the conventional coprecipitation approach to first create LDHs. However, prior to complete isolation, the solid is redispersed in acetone or methanol, both of which are miscible organic solvents in water and may alter the structure of the LDHs. There exists a possibility that certain precursors or reactants exhibit restricted solubility or stability in organic solvents, thereby restricting the range of LDH compositions that can be synthesized [30]. The AMOST method was used to synthesize the newly prepared Ni-Al-CO_3 LDH, which may serve as a viable candidate for CO_2 adsorption and provide a method for fine-tuning other LDH materials for a variety of applications.

3.4 Hydrothermal synthesis

The purity, crystallite size, coprecipitation, sol-gel, and urea hydrolysis-based synthesis of LDH can be improved through hydrothermal treatment. This method involves the formation of the desired LDH by physical mixing of their mixed metal oxide precursors. When MgO and Al_2O_3 are heated to 110°C, MgAl-based layered double hydroxides are formed. The modifications made in the washing phase prior to hydrothermal treatment aid in controlling the crystal morphologies, sizes, and stability of their aqueous solutions. By adopting a hydrothermal process, LDH enhances crystallinity and intercalation while also increasing particle size and morphological revolution. Regrettably, the lengthy synthesis time (ranging from hours to days) and the energy necessary to heat the solution are obstacles to the industrial application of hydrothermal treatment [31]. The energy consumption, long reaction times, maintaining uniform conditions and controlling the reaction kinetics at large scale, high equipment cost, and precursor stability toward harsh condition of hydrothermal treatment these are some of the challenges where research is going on to optimize hydrothermal treatment especially in carbon capture area. Despite these drawbacks, hydrothermal treatment remains a valuable method for LDH synthesis.

3.5 Coprecipitation

The coprecipitation method involves the use of a solvent-based base to facilitate the precipitation of metal in the form of hydroxide from a salt precursor. The prevailing and fundamental approach to synthesize LDHs with diverse compositions involves the coprecipitation of LDHs from aqueous solutions containing divalent and trivalent metal cations via the addition of bases such as aqueous NaOH (**Figure 2**). Particle aggregation was observed in the LDHs, which were typically finite particles with a broad particle size distribution that were produced by coprecipitation. To enhance the uniformity of the product and reduce the aggregation of particles, a range of post-synthetic aging methods and blending conditions have been implemented [32]. In contrast to materials acquired at variable pH, those produced by

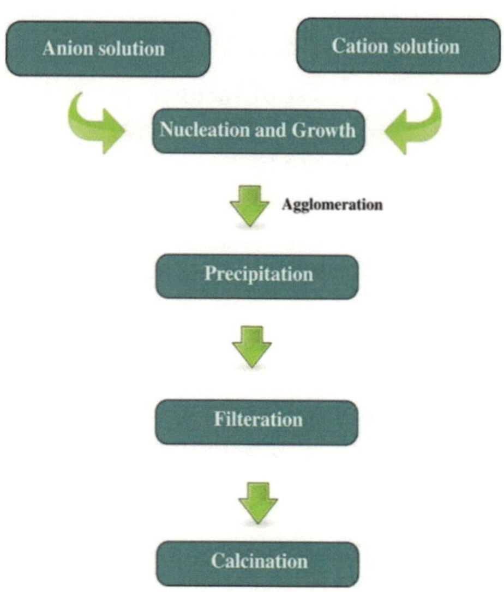

Figure 2.
Schematic of coprecipitation process.

coprecipitation at a constant pH exhibited an increased average pore diameter, decreased particle size, increased specific surface area, and enhanced crystallinity. LDH is frequently produced via the coprecipitation process, which employs metallic ions in alkaline media with a constant or rising pH. The resulting mixed oxides have unusual characteristics, such as restricted pore size distribution and large specific surface areas. LDHs possess distinct characteristics such as a well-defined layered structure, a specific anion exchange capacity, an adsorption capacity, and the ability to facilitate the movement of interlayer anions and water molecules [33]. When exposed to carbon dioxide and water, LDHs can reconstruct their structure during thermal degradation, forming stable and homogenous mixed oxides.

The benefits of the coprecipitation method are:

1. High yield.
2. High product purity.
3. High crystallinity.
4. Morphological revolution.
5. The absence of need to use organic solvents.
6. Mild reaction conditions.
7. Versatility in precursor selection.
8. Easily reproducible.
9. Low cost.

4. Modification of LDH

The aforementioned methods consist of various synthesis techniques, composition modification, morphology control, alkali metal enrichment of LDHs, and the fabrication of hybrid materials. Capabilities can be increased by incorporating amino acids or amino silanes into stable layers and reacting with hydroxyl groups on laminates to generate an amine loading. In contrast to the extensively researched LDH materials, there is a dearth of reported research on the adsorption of CO_2 by amine-grafted LDH.

4.1 Change in Mg/Al molar ratios

The highest-promising results, according to numerous studies, are observed at a Mg/Al molar ratio of 3:1 and temperatures ranging from 70 to 80°C. The CO_2 adsorption capacity of LDH increases with the presence of $NaNO_3$ in its structure, which results from an excess of magnesium nitrate; the maximum value observed was 9.27 mol/kg at 240°C and 1 atm CO_2 at a Mg/Al ratio of 20. A clear two-step process of CO_2 adsorption was identified in the Mg/Al molar ratio ranging from 12 to 30: an initial low CO_2 loading is followed by a progressive escalation to an exceedingly high CO_2 loading. The adsorption of CO_2 by LDH is influenced and facilitated by the concentration of $NaNO_3$ [34]. A similar-radius trivalent cation, such as Al^{3+}, enhances the positive charge density of the layer when it replaces the Mg^{2+} ion. To preserve electrical neutrality, interlayer sites, which are also involved in water crystallization, are occupied by a greater number of anions.

4.2 Modifying agents

One of the most crucial factors for CO_2 adsorbents is amine loading, which is the amount of amine in moles relative to the material's total weight. Appropriate Cu^{2+} ion addition can improve the surface morphology, free-amino functional groups, and amine loading of the materials, all of which can help with CO_2 adsorption and capture. The increase in Cu^{2+} content can enhance both chemical and physical adsorption, which is responsible for CO_2 adsorption. The rate of chemical adsorption reduced with increasing Cu^{2+} content. Using Cu-Mg-Al N as an active site, CO_2 can be photo-catalytically reduced to methanol, significantly improving CO_2 capture, and photocatalytic conversion [35].

4.3 Intercalation using nitrogen: containing anions

The presence of nitrogen anions can tune the LDH's affinity for CO_2, ensuring that it captures CO_2 while excluding other gases like nitrogen and oxygen. Nitrogen-containing anions, such as carbonate or bicarbonate, can be intercalated into LDHs to increase their CO_2 adsorption capacity. The total capture efficiency of LDHs is increased by the intercalated nitrogen anions, which offer more sites for CO_2 adsorption. The presence of carbonate and anions derived from urea was observed during the synthesis of layered double hydroxide using the urea hydrolysis method. According to an elemental study of these LDHs, intercalation of nitrogen anions between LDHs may boost CO_2 capture capacity. The results obtained indicate that the urea-derived anions prefer to intercalate at low temperatures and concentrations of aluminum, whereas carbonate anions predominate in samples with high concentrations of aluminum and are prepared at a high synthesis temperature (100°C) [36]. LDH composed of Mg-Al has been the most prevalent due to its minimal preparation cost and high adsorption capacity. On account of its distinct morphology and layered structure, it is amenable to alterations that would increase its CO_2 adsorption capacity. LDHs experience interlayer water dehydration, interlayer CO_3^{2-} group release, and dehydroxylation of layered OH groups across a range of temperatures. This results in the formation of an alkali-doped amorphous solid oxide with greater stability at elevated temperatures and a greater surface area [5].

4.4 Doping with alkali metal

Alkali metals help LDHs enhance their capacity to adsorb CO. Doping K_2CO_3 on LDHs is thought to alter the chemical and physical characteristics of LDHs, increasing their capacity to adsorb CO. Furthermore, even if brunauer-emmett-teller (BET) area and pore volume are reduced, it is hypothesized that K_2CO_3 impregnation boosts the adsorption capacity of LDHs by increasing the number of active sites on the surface [37, 38].

5. Conclusion

The most researched materials for low-temperature applications (T < 200°C) include activated carbon, zeolites, and MOFs; hydrotalcites and oxides are typically used for intermediate and high-temperature applications (200 < T < 400°C and T > 400°C), respectively. Activated carbons are inexpensive and simple to functionalize materials. In comparison to carbons and zeolites, MOFs typically have a larger capacity for CO_2 adsorption, along with strong selectivity and easy structural tunability. Mesoporous silicas are appealing materials that have a lot of meso- and micropores as well as a lot of reactive Si—OH groups on the pore surface. These groups provide adsorption sites as well as appropriate centers for functionalization [14]. Materials based on ACs have been used for the adsorption and separation of CO_2 in humid environments. It has also been noted that the preparation methods have a significant impact on the pore size and, consequently, the structure's active surface area [16]. It has been discovered that several techniques, including the activation process, surface modifications, and the creation of AC-based composites with a variety of additives, including reduced graphene oxide, metal-organic frameworks, and zeolites, can optimize the surface morphology of activated carbon. Even though ACs have a high absorption of CO_2 and an excellent selectivity of up to 48, there is still an opportunity for improvement in order to guarantee that high levels of recovery and purity of the product gases may be reached, meeting industrial standards. When ACs are run at high temperatures and their gases are subjected to high moisture content, they are also susceptible to poor adsorption performance. Therefore, it is imperative to develop innovative activated carbon material designs that can survive these conditions in order to achieve greater CO_2 extraction from gas mixes, flue gas, or biogas [17, 19].

Porosity, high surface area, functionality, high regenerative ability, outstanding electrical properties, and high gas storage capacity are some of the special qualities of carbon based materials (CBMs). The carbon-derived materials utilized for CO_2 collection are organized and functionalized as graphene, membranes, aerogels, CNTs, and fullerene according to their size, shape, and dimensionality [39].

One of the best candidates to be employed for CO_2 capture is mixed metal oxide (MMO) adsorbents at moderate temperatures (200–450°C). They are inexpensive, simple to synthesize, stable, and react well to both high pressures and the presence of steam. Their biggest flaw is their adsorption capacity, which can be strengthened by precisely regulating the adsorption and synthesis conditions. Particularly, the graphene oxide/LDH combination not only boosts the samples' adsorption capacity but also gives the structure mechanical stability that lasts for multiple cycles [25].

In the area of photocatalysis, LDHs have more room for growth. This is primarily because their structure's flexibility promotes the advantageous development of their photoelectric qualities. The distinct stratified configuration offers optimal circumstances for the even dispersion of reactive sites. As a result, research on structural modifications for photocatalytic CO_2 reduction enhancement is essential to understanding how LDHs are used in photocatalysis [40].

A number of critical considerations must be made for industrial applications, including the precalcination time, the heating rate, the cost analysis, and the adsorbent's lifespan under actual regeneration conditions. Aside from the co-adsorption of CO_2 and water, information regarding the co-adsorption of other gases in LDHs is limited. Studying parameters pertinent to industrial operating conditions can be done when evaluating LDHs for CO_2 separations from other gases in a manufacturing environment.

Subsequent investigations may concentrate on refining LDH structures to augment their ability to adsorb CO_2, investigate adjustments in LDH compositions or integrate functional groups to enhance selectivity and diminish energy demands for the separation process. This includes modifying LDHs' surface area, layer spacing,

DOI: http://dx.doi.org/10.5772/intechopen.1005188

and composition to optimize how well they interact with carbon dioxide molecules. Due to their shorter synthesis pathways and reduced chemical consumption, direct modification techniques are typically selected. More research is required in this area because, despite the abundance of reports on the subject of LDH-based adsorbents and photocatalysts, most of the studies are restricted to batch-scale operations and are not substantially developed at a commercial level.

There is an urgent requirement to assess the tangible process obstacles linked to the implementation of LDH, including logistics and management, cost-effective synthesis and regeneration on a larger scale, and process development. Research should explore the development and application of novel adsorbent materials beyond conventional activated carbons and zeolites. Additionally, research should be conducted on the hazard assessment of LDH and LDH-containing hybrids, as well as their impact on the environment and human existence in a broader context.

A promising technology for reducing carbon dioxide emissions from power facilities and industrial processes is adsorbent-based carbon capture. The performance and economic viability of adsorption-based carbon capture in comparison to alternative capture technologies, such as membrane separation or amine-based absorption, must be assessed in order to determine the advantages and disadvantages of each strategy. Innovative techniques, such as solar- or microwave-assisted regeneration, can be investigated in order to raise the carbon capture process's total energy efficiency.

The use of these adsorbents for large-scale CO_2 adsorption is limited by their low production of adsorbent materials, poor stability and long-term durability over multiple adsorption-desorption cycles, expensive adsorption systems, high feed gas volumetric flow rate, high accommodation requirement for adsorbate in the feed during adsorption, duration of gas flow path switching between adsorption and regeneration, high capture cost in post-combustion, large CO_2 separation, and the presence of water (approximately 7%) and impurities in the flue gas. Investigating potential uses for captured CO_2, such as in synthetic fuels or chemical production, can enhance the economic viability of adsorbent-based capture.

Author details

Himangi Neve*, Deven Kadam, Chinmay Sadadekar, Vinayak Wadgaonkar, Siraj Bhatkar and Prasad Wadekar
Dr. Vishwanath Karad MIT World Peace University, Pune, India

*Address all correspondence to: himangi.neve@mitwpu.edu.in

References

[1] Wang Q, Tay HH, Zhong Z, Luo J, Borgna A. Synthesis of high-temperature CO_2 adsorbents from organo-layered double hydroxides with markedly improved CO_2 capture capacity. Energy & Environmental Science. 2012;**5**:7526. DOI: 10.1039/c2ee21409a

[2] Ruether JA. FETC programs for reducing greenhouse gas emissions. In: Report DOE/FETC-98/1058; U.S. Department of Energy, U.S. Washington, DC: Government Publishing Office; 1999

[3] Wang J, Huang L, Gao Y, Yang R, Zhang Z, Guo Z, et al. A simple and reliable method for determining the delamination degree of nitrate and glycine intercalated LDHs in formamide. Chemical Communications. 2014;**50**:10130. DOI: 10.1039/C4CC05015K

[4] Wang H, Liu Y, Li J. Designer metal-organic frameworks for size-exclusion-based hydrocarbon separations: Progress and challenges. Advanced Materials. 2020;**32**:Article 2002603(44). DOI: 10.1002/adma.202002603

[5] Elkartehi ME, Mahmoud R, Zaher A. LDH nanocubes synthesized with zeolite templates and their high performance as adsorbents. Nanomaterials (Basel). 2021;**11**(12):3315. DOI: 10.3390/nano11123315

[6] Megias-Sayago C, Bingre R, Huang L, Lutzweiler G, Wang Q, et al. CO_2 adsorption capacities in zeolites and layered double hydroxide materials. Frontiers in Chemistry. 2019;**7**:1

[7] Yang ST, Kim J, Ahn WS. CO_2 adsorption over ion-exchanged zeolite beta with alkali and alkaline earth metal ions. Microporous and Mesoporous Materials. 2010;**135**:90-94

[8] Cavallo M, Dosa M, Porcaro N, et al. Shaped natural and synthetic zeolites for CO_2 capture in a wide temperature range. Journal of CO_2 Utilization. 2023;**67**:102335. DOI: 10.1016/j.jcou.2022.102335

[9] Zhang Z, Yao Z-Z, Xiang S, Chen B. Perspective of microporous metal–organic frameworks for CO_2 capture and separation. Energy & Environmental Science. 2014;**7**:2868-2899

[10] Ma XC, Li LQ, Chen RF, Wang CH, Li HL, Wang SB. Heteroatom-doped nanoporous carbon derived from MOF-5 for CO_2 capture. Applied Surface Science. 2018;**435**:494-502

[11] Demir H, Aksu G, Gulbalkan H, et al. MOF membranes for CO_2 capture: Past, present and future. Carbon Capture Science and Technology. 2022;**2**:100026. DOI: 10.1016/j.ccst.2021.100026

[12] Mello MR, Phanon D, Silveira GQ, Llewellyn PL, Ronconi CM. Amine-modified MCM-41 mesoporous silica for carbon dioxide capture. Microporous and Mesoporous Materials. 2011;**143**:174

[13] Herawan SG, Hadi MS, Ayob MR, Putra A. Characterization of activated carbons from oil-palm shell by CO_2 activation with no holding carbonization temperature. Scientific World Journal. 2013;**2013**:6. Article ID 624865. DOI: 10.1155/2013/624865

[14] Tumurbaatar O, Popova M, Mitova V, Shestakova P, Koseva N. Engineering of silica mesoporous materials for CO_2 adsorption. Materials. 2023;**16**(11):4179. DOI: 10.3390/ma16114179

[15] Abuelnoor N, AlHajaj A, Khaleel M, Vega LF, Abu-Zahra MRM. Activated carbons from biomass-based sources for CO_2 capture applications. Chemosphere. 2021;**282**:131111. DOI: 10.1016/j.chemosphere.2021.131111

[16] Morali U, Demiral H, Sensoz S. Synthesis of carbon molecular sieve for carbon dioxide adsorption: Chemical vapor deposition combined with Taguchi design of experiment method. Powder Technology. 2019;**355**:716-726. DOI: 10.1016/j.powtec.2019.07.101

[17] Abd AA, Othman MR, Kim J. A review on application of activated carbons for carbon dioxide capture: Present performance, preparation, and surface modification for further improvement. Environmental Science and Pollution Research. 2021;**28**:43329-43364. DOI: 10.1007/s11356-021-15121-9

[18] Park J, Cho SY, Jung M, et al. Efficient synthetic approach for nanoporous adsorbents capable of pre-and post-combustion CO_2 capture and selective gas separation. Journal of CO_2 Utilization. 2021;**45**:101404. DOI: 10.1016/j.jcou.2020.101404

[19] Hosseini S, Bayesti I, Marahel E, et al. Adsorption of carbon dioxide using activated carbon impregnated with Cu promoted by zinc. Journal of the Taiwan Institute of Chemical Engineers. 2015;**52**:109-117. DOI: 10.1016/j.jtice.2015.02.015

[20] Pal A, Uddin K, Rocky KA, et al. CO_2 adsorption onto activated carbon–graphene composite for cooling applications. International Journal of Refrigeration. 2019;**106**:558-569. DOI: 10.1016/j.ijrefrig.2019.04.022

[21] Zhao Y, Ding H, Zhong Q. Preparation and characterization of aminated graphite oxide for CO_2 capture. Applied Surface Science. 2012;**258**:4301

[22] Meng LY, Park SJ. Effect of exfoliation temperature on carbon dioxide capture of graphene nanoplates. Journal of Colloid and Interface Science. 2012;**386**:285-290. DOI: 10.1016/j.jcis.2012.07.025

[23] Kuang Y, Zhao L, Zhang S, Zhang F, Dong M, Xu S. Morphologies, preparations and applications of layered double hydroxide micro-/nanostructures. Materials (Basel). 2010;**3**(12):5220-5235. DOI: 10.3390/ma3125220

[24] Yang Z-Z, Wei J-J, Zeng G-M, Zhang H-Q, Tan X-F, Ma C, et al. A review on strategies to LDH-based materials to improve adsorption capacity and photoreduction efficiency for CO_2. Coordination Chemistry Reviews. 2019;**386**:154-182. ISSN 0010-8545. DOI: 10.1016/j.ccr.2019.01.018

[25] Santamaría L, Korili SA, Gil A. Layered double hydroxides for CO_2 adsorption at moderate temperatures: Synthesis and amelioration strategies. Chemical Engineering Journal. 2023;**455**:140551. ISSN 1385-8947. DOI: 10.1016/j.cej.2022.140551

[26] Kong TT, Huang J, Jia XG, et al. Synthesis and optimization of Ti/Li/Al ternary layered double hydroxides for efficient photocatalytic reduction of CO_2 to CH_4. Scientific Reports. 2019;**9**:5659. DOI: 10.1038/s41598-019-41979-4

[27] Tang LQ et al. ZnxCd1−xS tunable band structure-directing photocatalytic activity and selectivity of visible-light reduction of CO_2 into liquid solar fuels. Nanotechnology. 2018;**29**:064003

[28] Joos L, Swisher JA, Smit B. Molecular simulation study of the competitive

adsorption of H_2O and CO_2 in Zeolite 13X. Langmuir. 2013;**29**:15936-15942

[29] Sokol D, Vieira DEL, Zarkov A, et al. Sonication accelerated formation of Mg-Al-phosphate layered double hydroxide via sol-gel prepared mixed metal oxides. Scientific Reports. 2019;**9**:10419. DOI: 10.1038/s41598-019-46910-5

[30] Zhi Ping X, Qing (Max) Lu G. Hydrothermal synthesis of layered double hydroxides (LDHs) from mixed MgO and Al_2O_3: LDH formation mechanism. Chemistry of Materials. 2005;**17**(5):1055-1062

[31] Ashik UPM, Kudo S, Hayashi J-I. An Overview of Metal Oxide Nanostructures. In: Bhagyaraj SM, Oluwafemi OS, Kalarikkal N, Thomas S, editors. Micro and Nano Technologies, Synthesis of Inorganic Nanomaterials. Woodhead Publishing; 2018. pp. 19-57. ISBN 9780081019757. DOI: 10.1016/B978-0-08-101975-7.00002-6

[32] Trifiro F, Vaccari A. Handbook of Layered Materials. New York, NY, USA: Marcel Dekker; 2004. p. 251

[33] Elhenawy SEM, Khraisheh M, AlMomani F, Walker G. Metal-organic frameworks as a platform for CO_2 capture and chemical processes: Adsorption, membrane separation, catalytic-conversion, and electrochemical reduction of CO_2. Catalysts. 2020;**10**:1293. DOI: 10.3390/catal10111293

[34] Wang J, Stevens LA, Drage TC, Wood J. Preparation and CO_2 adsorption of amine modified Mg-Al LDH via exfoliation route. Chemical Engineering Science. 2012;**68**(1):424-431. ISSN 00092509. DOI: 10.1016/j.ces.2011.09.052

[35] Faramawy S, Zaki T, Sakr AA-E, Saber O, Aboul-Gheit AK, Hassan SA. The activity of Mg-Al layered double hydroxides intercalated with nitrogen-containing anions towards the removal of carbon dioxide from natural gas. Journal of Natural Gas Science and Engineering. 2018;**54**:72-82.ISSN 1875-5100. DOI: 10.1016/j.jngse.2018.04.002

[36] Sharma U, Tyagi B, Jasra RV. Synthesis and characterization of Mg-Al-CO_3 layered double hydroxide for CO_2 adsorption. Industrial and Engineering Chemistry Research. 2008;**47**(23):9588-9595

[37] Yang JI, Kim JN. Hydrotalcites for adsorption of CO_2 at high temperature. Korean Journal of Chemical Engineering. 2006;**23**:77-80. DOI: 10.1007/BF02705695

[38] Wang Q, Tay HH, Ng DJW, Chen L, Liu Y, Chang J, et al. The effect of trivalent cations on the performance of Mg-M-CO_3 layered double hydroxides for high-temperature CO_2 capture. ChemSusChem. 2010;**3**:965-973

[39] Zaker A, Hammouda SB, Sun J, Wang X, Li X, Chen Z. Carbon-based materials for CO_2 capture: Their production, modification and performance. Journal of Environmental Chemical Engineering. 2023;**11**(3):109741. ISSN 2213-3437. DOI: 10.1016/j.jece.2023.109741

[40] Ding G et al. Layered double hydroxides and their composites as high-performance photocatalysts for CO_2 reduction. EES Catalysis. 2023;**1**(4):369-391. DOI: 10.1039/D3EY00080J